Tap Water as a Hydraulic Pressure Medium

Tap Water as a Hydraulic Pressure Medium

Erik Trostmann
Birkeroed, Denmark

Bo Frølund
Bo Højris Olesen
Danish Technological Institute
Århus, Denmark

Bjarne Hilbrecht
Danfoss Water Hydraulics
Nordborg, Denmark

CRC Press is an imprint of the
Taylor & Francis Group, an **informa** business

CRC Press
Taylor & Francis Group
6000 Broken Sound Parkway NW, Suite 300
Boca Raton, FL 33487-2742

First issued in paperback 2019

ISBN-13: 978-0-8247-0505-3 (hbk)
ISBN-13: 978-0-367-39790-6 (pbk)

Visit the Taylor & Francis Web site at
http://www.taylorandfrancis.com

and the CRC Press Web site at
http://www.crcpress.com

Foreword

I became excited after reading this book, and highly recommend that it be read by those in the hydraulic industry and government agencies. *Tap Water as a Hydraulic Pressure Medium* is a needed reference that dispels many of the judgments made by those with an inadequate background in water hydraulics. This book is a must, as water hydraulics will be significant in the mobile and industrial sectors in the next five years.

The book discusses applications that are based on the advantages of water hydraulics, such as environmental friendliness, more efficient and responsive systems, nonflammability, and a high heat rejection rate. It is invaluable for engineers as it answers questions about solving extreme system temperature conditions in water with additives such as biodegradable liquids. Water constituents that lead to chemical bond and system material interactions, as well as bacteria-growth concerns, are addressed in the chapter on water chemistry.

The water properties table compares water and common oils, thus allowing projection of many future applications. Some of these can only be found in this text, including viscosity, vapor pressure, speed of sound, bulk modulus, thermal conductivity, specific heat, density, and surface tension.

Professor Gary Krutz, Ph.D., P.E.
Director, Electrohydraulics Center
Purdue University
West Lafayette, Indiana

Preface

From a historic perspective, water was the first hydraulic medium ever to be used. It was not until the year 1900 that mineral oil–based media were introduced, and they have since almost completely taken the place of water in many technical applications. However, oil grows old and degrades with use and must therefore be renewed and incinerated. In addition, oil can cause fires, and leaks in systems and spillage due to handling cause pollution. Moreover, hydraulic oils must be procured by just a few producers worldwide. Increasing environmental and safety demands and a general need for a cleaner technology has revived the thought of water as a medium, which would alleviate the above-mentioned drawbacks. Technological development over the past decades has created materials, surface treatments, production processes, and calculation possibilities that are now making possible production of technically competitive water hydraulic systems. Water hydraulics has been reborn and deserves support to spread into those areas of use where its advantages lie.

The intention of *Tap Water as a Hydraulic Pressure Medium* has been to create a comprehensive encyclopedia providing the knowledge one might need. The physics, chemistry, and microbiology of water are described. The chapter "Water Treatment" offers practical recommendations from real life for keeping installations clean and fit for special demands. Microbiologically influenced corrosion of metal is the focus of the chapter "Water Microbiology".

The book may be used for development of components and for the drafting, building, operation and maintenance of water hydraulic systems; the technical depth of the book has been adjusted accordingly. Information about water that is of no relevance to hydraulics has been omitted, whereas background information necessary for reaching an understanding of relevant conditions has been included. Adding to the usefulness of the book, examples have been woven into the text.

We imagine the main user of the book to be an engineer, particularly one with a technical background in mechanics. Even though chemistry and microbiology are also emphasised in the book, it has been written primarily for non-chemists and non-microbiologists. We have attempted to describe the interrelationships of water properties to make the qualities of water visible; for example, the text concerning the physics of water includes comparisons with hydraulic oil. We hope that the book will serve as a basis for elaboration of concise information sheets and guidelines that specific user groups—e.g., system designers, fitters or operational staff—may need.

The initiative for this book can be traced back to 1989, when Danfoss A/S decided to direct efforts toward water hydraulics and initiated development of water hydraulic components. The Foreword of the book *Water Hydraulics Control Technology* by Erik Trostmann, Technical University of Denmark, tells this story. The current volume should be seen as amplification of a sub-topic of the earlier book. In 1997, the Danfoss initiative was followed up by the launch of the EUREKA project entitled HYDRA with several German and Danish project partners. HYDRA consists of several sub-projects, including "Properties of Tap Water as a Hydraulic Pressure Medium". This book is the result of this sub-project.

Knut Meyer
Project Manager, HYDRA
Senior Engineer
Institute for Product Development
Technical University of Denmark

Acknowledgement

A piece of work such as this book has many contributors, of course, who through personal commitment and enthusiasm create the basis for the accomplishment and success of the project. I should especially like to thank Professor Erik Trostmann as the initiator of the Danish contribution to the EUREKA project HYDRA, as a visionary and committed spokesman for water hydraulics in areas where its advantages have possible applications, and for his persistent effort to keep the flag flying, technically speaking.

We thank EUREKA/Denmark for providing the financial basis for the deveopment of this book within the responsibility of the Department of Control and Engineering Design at the Technical University of Denmark (DTU), thus allowing this initiative to bear fruit. The openness and commitment of Danfoss A/S have contributed substantially to the quality of the result. From Germany, Sascha Brauer, INNOSYS GmbH & Co. KG, Bochum (D), and Andreas Alker, Danfoss GmbH, Offenbach (D) have contributed with their knowledge of water as a hydraulic pressure medium. Esmaeil Varandili, of the Department of Control and Engineering Design, DTU, made a remarkable effort in assembling and structuring information for this book.

Although it was not planned at the start of the HYDRA project, we later found it reasonable to present the results of this sub-project as a book, published by Marcel Dekker, Inc., which also published Prof. Erik Trostmann's *Water Hydraulic Control Technology*. The following partners of the EUREKA project HYDRA together with the Danish EUREKA secretariat and two universities, which are working with Water Hydraulics, established the financial basis for the book:

- CWO Semco Maritime, Svendborgvej 253, DK-5260 Odense S
- Danfoss A/S, DK-6430 Nordborg

- Department of Control and Engineering Design, Technical University of Denmark, DK-2800 Kgs. Lyngby
- Dr. Breit GmbH, Carl-Zeiss-Strasse 25, D-42579 Heiligenhaus
- Erhvervsfremme Styrelsen, Erhvervsministeriet, Langelinie Allé 17, DK-2100 Copenhagen
- Freudenberg Dichtungs- und Schwingungstechnik KG, D-69465 Weinheim
- Institute of Hydraulics and Automation, Tampere University of Technology, FIN-33101 Tampere
- Joseph Vögele AG, Neckarauer Strasse 168-228, D-68146 Mannheim

Last, but not least, I thank the authors for making the role of Project Manager so easy.

Knut Meyer

Contents

1 Using water in hydraulic systems

Per Sørensen
Danfoss A/S

1.1 Summary

Water hydraulics has been in existence for just over 200 years and is now beginning to be fully exploited, with a number of new technologies and products having been developed.

This article assesses the performance of water hydraulics in comparison with oil hydraulics, electric and pneumatic systems. It briefly discusses the types of water that can be used, outlines the engineering challenges in developing water hydraulic components, and assesses a water hydraulic user's view on the advantages gained by "going water".

1.2 Background/history

Between 1800 and 1820, the first significant water hydraulic systems were developed in the UK and by 1868 - in the mid-Victorian era - a 65 mile ring main system was established in London. This system later came to be run by the London Hydraulic Power Company.

The system drew water from the River Thames and Regents Canal by means of several pumping stations. It produced a minimum guaranteed pressure of approx 50 bar.

This system supplied energy for lifts and cranes around London and in case of fire, the system supplied water for fire fighting.

In 1894 Tower Bridge was opened to road and river traffic. The bridge, designed by Horace Jones and Sir John Wolfe Barry, was solidly built and the 52 bar water hydraulic system installed in it remained in use until July 1971, when a contract to modernise it and switch to an oil system was won by the Cleveland Bridge and Engineering Company.

Water hydraulics has also been used in many other European countries. One of the most famous water hydraulic applications can be found in the Eiffel Tower in Paris. In 1887, when the tower was built, a water hydraulic lift was installed. This lift is still driven by water.

In Germany water hydraulics has also been in use in many cases. In Cologne they have just recently modernised a water hydraulically operated bridge over 100 years old, with new developed modern water hydraulic components.

To add to the burdens of water hydraulics, by 1900 mineral oil had become available in larger quantities and 20 years later, in 1920, compact hydraulic systems began to be developed.

However, as with all product development stories - and particularly those stretching over a long time - what goes around comes around.

By 1940 there was a growing demand for non-flammable fluids - a demand which had always been met by water hydraulics, of course. The environmental card then began to be played and by 1988, Germany and Sweden in particular were expressing serious interest in hydraulic systems based on biologically degradable fluids. The same year, one of the world's largest hydraulic product manufacturers began an R&D project which eventually led to a new generation of modern water hydraulic products.

Of course there are other players in the game - at this time approximately 12 companies are manufacturing water hydraulic components. However, in 1994 Danfoss A/S became the first company in the world to introduce a complete range of water hydraulic products with the same characteristics as oil hydraulics.

As the modern range of water hydraulic systems produces the same results and performance as an oil filled system, it delivers all the advantages of oil, with none of the disadvantages. In addition, the products are corrosion resistant, reliable, easy to clean and have only a few wearing parts.

1.3 Water

In many ways, as engineers we think of water as H_2O. Perhaps we also think of the EU drinking water directive 80/778. But we also think of the solid particles it often contains, its soluble substances, the bacteria it carries, the hydrogen-ion concentration, and its level of chloride, calcium and magnesium.

In developing the new generation of water hydraulic systems, water presented a number of inherent problems:

- it can leave lime deposits;
- it can erode and corrode the material through which it runs; and
- it can carry bacteria into the system where they might grow.

These topics will be dealt with later in this book.

In considering 'water' one must also consider water anti-freeze additives - something that is important in applications exposed to frost (below zero °C). More specifically, the systems are designed to operate at temperatures of 3–50°C.

As a general, but rather unscientific rule for water quality acceptable as hydraulic medium, we say "If you can drink it, we can run it". However, any water type must be passed through a 10 Micron absolute filter before used as hydraulic medium in order to catch any possible sediments.

1.4 Water versus oil and others

Some general statements about oil form a compelling argument as to why water hydraulic systems are so popular and important:

"One drop of oil spoils 150 litres of drinking water."
"The cost of cleaning 1 ton of soil after an oil spill, depending on the degree of pollution, can be as much as USD 2,000."

If we look at HLP, the price of one litre of HLP-oil is about USD 1. Whilst not free, the price of one litre of water is around 1/500 Cent; 50,000 times cheaper. However, mineral oil is not the only hydraulic medium currently in use. There are biological oils, such as rape seed. There are also the water/oil emulsions HFA, HFB and HFC.

HFA, HFB and HFC all have reduced fire risks when compared to mineral oil, but they are still environmentally damaging. Bio-degradable oil, which will not damage the environment, is a potential fire hazard and still slow to degrade.

As a generalisation, it is fair to say that no medium offers all the advantages of water. In addition, water is a low cost medium in terms of purchase, storage and disposal. It also offers improved system stiffness due to its lower compressibility. These aspects are dealt with in the chapters to follow.

1.4.1 Water hydraulics vs. electrical drives

Unlike most common electrical systems, water hydraulics can withstand all common forms of cleaning - something particularly useful in the food, chemical and petrochemical industries - and not something of which electrical systems are capable, without major modification.

In a water-based system the heat generated in pumps, motors, valves and piping is transmitted into the hydraulic liquid, which can be easily and cheaply cooled - away from temperature-critical areas. Speed - rotational and linear - can be infinitely regulated without the need for expensive additional components.

One of the main problems with electric motors are that they need cooling. In the food industry, applications also demand hygiene, water proofing of the components and/or constant ambient temperature. As a result, motors are sometimes built into stainless steel enclosures, which have to be dismantled and cleaned as well as ventilated - increasing servicing requirements.

Furthermore, electric motors are often positioned away from the wet processing zone, - which means that power has to be transferred via shafts, chains, gear-wheels, etc - creating possible lubrication, screening and hygiene problems - and, of course, calling for more space and costs.

To summarise, when compared to electrical systems, water hydraulics have the following advantages:

- very compact components
- minimal heat generation
- they are sanitary and easy to clean
- they do not have the potential to electrocute their operators
- encapsulation is not necessary, and they can operate directly in wet zones

1.4.2 Water hydraulics vs. pneumatics

Pneumatic systems eliminate the risk of oil contamination. However, they do have their own problems. When compared to pneumatics, water hydraulics has significant efficiency and regulation advantages. Also, the volume of water hardly changes under pressure, giving good positioning characteristics; its low viscosity means low pressure drop, and the higher pressure level (140 bar) results in compact component design.

Pneumatics does not have the same risk of food contamination as oil (although air exhaust often contains oil, fungi, germs, etc). Pneumatics has a poor efficiency and some hygiene and ecological problems. Pneumatic systems can leak up to 25% of their compressed air into production areas and they are sometimes 'springy', with unstable movement and poor positioning, requiring liquid-filled hydraulic stabilizing cylinders.

Finally, their relatively low operation pressure means motors and cylinders often have to be significantly larger than corresponding hydraulic equipment.

Overall, pneumatics tend to be inefficient, with possible power losses of up to 85-90% when compressors, air-drying equipment, pipes, valves, motors and cylinders are included. This can mean applying up to 40–50 kW of energy to extract a useful output of 3.5 kW.

By substituting pneumatic motors on rotating meat saws with water hydraulic motors, Danish slaughterhouses have achieved energy savings up to 88%! This corresponds to over USD 1,250 per year per saw installation.

1.5 Water - the engineering challenges

Nothing in life is ever simple. Whilst water is the obvious answer to hydraulic requirements, in terms of the environment and usage in hazardous environments, it does pose some engineering challenges.

These challenges are hinted at by the viscosity figures. Water has a viscosity of 1 centistoke, and oil has a viscosity of 37 centistokes. A challenge is therefore in sealing water-based hydraulic systems.

When solving the problem posed by water's viscosity, special attention should be paid to possible pump leakages, as an internal leakage could reduce the flow of the pump - and thereby its volumetric efficiency.

Using experience, gained from producing approximately 10,000 hydraulic components per day, Danfoss whittled down parts tolerances to a point where volumetric efficiencies became over 96% at maximum rated pressure. External tightness has been ensured by using special O-rings and a specially designed mechanical axial seal around the output shaft.

In addition, there are the potential corrosion, lubrication, erosion and cavitation challenges inherent in water as mentioned earlier.

Water hydraulics has developed around two types of core material - plastics and ceramics. In 1988 Danfoss A/S, for instance, decided to go down the plastics route. Although the more expensive option, plastics are more flexible - both in the way they can be processed and in their application.

However, to solve the corrosion problems, it was necessary to develop a special plastic based around a polymer compound. The rest of the system is made from stainless steel, of adequate qualities.

To solve the lubrication problems all bearing surfaces are formed as hydrodynamic surfaces, except for the connection between swash-plate and the side shoes, which are formed as specially developed hydrostatic bearings.

Owing to the low viscosity and the high pressure differentials, any fissure in the pump will cause a very high water speed over the fissure. In theory this can lead to erosion on the surfaces, especially if the water contains any small solid particles. This particular engineering problem was solved by using specially hard components and incorporating smooth flow paths.

As an illustration of the detail involved in developing the new component designs, many (over 30) patents have been taken out by only one manufacturer.

Due to the potential erosion and lubrication problems, component manufacture has to be significantly more precise than with oil-filled systems.

The last major problem is caused by cavitation. Water has a lower boiling point than oil, and steam may form at temperatures as low as 55°C on pump suction lines. In theory, this means that the steam can cause surface cavitation as it travels through the system and explosively expands when it reaches a pump, potentially causing it to fail.

On the other hand, the lower viscosity of the water means lower pressure losses in system. As a result, cavitation has not been found a problem in practice provided that maximum temperature is not exceeded.

1.6 Applications

Industrial applications of water hydraulics today are typical in the following industries:

- food processing (- agitators, belt conveyors, saws, rotating knives)
- pharmaceutical
- chemical
- nuclear power and treatment
- water-mist (fire fighting, moistening, lumber drying)
- off-shore
- mobile (lorries, boats, hovercrafts, etc)
- railway brakes

1.7 Statements

Let a daily user of water hydraulics express his experience with water as a hydraulic medium:

Statements made by Lars Thulin, Production Manager with Renova A/B, the Municipal Sanitation Department in Gothenburg, Sweden:

"Water Hydraulics - The best alternative to oil hydraulics - when health, environment and economics must be considered".

1.7.1 Hydraulics oil – a large latent environmental problem

Hydraulics oil
40,000 tons leak into the Swedish nature every year
Limited natural resource
In a large Danish city, municipal vehicles use 18,000 litres annually
May cause diseases (a.o. allergy)
Is toxic and inflammable
Is an indirect reason for traffic accidents
Causes damage to streets and roads
Presents contamination risk in goods production. Thus unwanted in many industries
Is hazardous waste

Water
Our most environmentally friendly fluid.
Unlimited natural resource (still!).
Entirely non-toxic.
Low-priced.

1.7.2 Handling hydraulics oil and water

Hydraulics oil to be
Purchased
Handled
Stored
Used
Transported and received as hazardous waste
Stored again
Transported
Disposed of

Water to be
Tapped
Filled
Used
Discharged
.... poured into the sewer

1.7.3 Water hydraulics - costs and winners

Costs	
Water	0.01 DKK/l
Increased price of technology	~+20%
Service cost	lower
Absence due to illness	none
Indirect traffic accidents	none
Fire hazard	none
Road surface costs	none
Environmental costs	none
Product contamination	none
Disposal costs	none

Winners
Environment
Endusers
Enginemen
Maintenance personnel
.... Next generation

1.7.4 Water hydraulics – something to think about!

- Ordinary tap water and propylene glycol for frost protection
- High quality of the hydraulics system (consisting of ceramics, plastics, and stainless steel)
- No allergy risk (producers, distributors, machine and service personnel, etc)
- No need for protection equipment
- No need for special workshop – repairable on site
- Lower service cost
- Hydraulics water is clean "waste"
- No "oil spillage" and no costs for traffic warnings and oil cleaning – inside as well as outside
- Water does not damage food
- Water does not damage roads and road surfaces
- No need for destruction (cost of destruction = new price of hydraulics oil)

1.8 Conclusion/future

A tendency in the application of water hydraulics is that more and more industrial sectors show interest in applying the technology. It is seen that the systems are becoming more and more sophisticated, which again calls for further development of components and special solutions, as well as combination of several manufacturers' components.

References

Higgins, Mike, *"Water hydraulics - a technology comes of age"*, Seminar at Danfoss A/S, 1995.

Trostmann, Erik, *"Water Hydraulics Control Technology"*, Marcel Dekker, Inc., 1996.

DEFU, "Energy optimized hydraulics (2)", DTU, (January 2000).

Thulin, Lars, *"Statements of using water as hydraulic medium"*, Renova A/B, Gothenburg, Sweden, (1999).

Sørensen, Per, "News and trends by the industrial application of water hydraulics", SICFP'99, Tampere, Finland, (May 1999).

2 Introduction to water

Bo Højris Olesen
Danish Technological Institute

If somebody asked you:
"What is water?"
How would you respond?
It's a rather simple question,
but would you be able to answer it...correctly?
Mistakenly, you might reply:
"Water is just water"
revealing that you, like the rest of us, take water for granted,
failing to realise how important water really is.

This introduction will not try to give the correct answer to the question above, nor will it try to cover all we know about water today. Instead, this chapter will present you with bits and pieces of "common knowledge" describing some of the many aspects related to water. Hopefully, this will enable you to see the rest of this book, the physics, the chemistry, and the microbiology of waters, in a broader perspective.

2.1 Fount and origin of life

Looking at water, your first thought may be that it's the simplest thing around. But, has it ever occurred to you that in fact water is vital for all life on earth? Where there is water, there is life. Without water there would be no life at all. Water is the one thing that brought our planet to life.

Take a good look at yourself. What do you need to really just survive? Depending on your way of life the answer could be almost anything, but it would almost certainly contain food and water. Though human beings can survive without food for months, they can only stay alive for about a week without water. In order to maintain good health, people need about 2.5 litres of water each day (from drinking or eating).

Did you know...?
All living things contain large amounts of water. Here are some examples:

Human:	60% water	Egg:	74% water
Elephant:	70% water	Earthworm:	80% water
Tree:	70% water	Watermelon:	92% water
Meat:	70% water	Tomato:	95% water

Water is of major importance to all living organisms due to its unique qualities and properties. The cells that make up the human body, or for that matter any other living plant or animal, contain a large amount of water. The human body, for example, contains about 60% water. The ability of water to dissolve almost anything allows our cells to use valuable nutrients, minerals, and chemicals. The high surface tension of water (the fact that water sticks together) plays an important part in our body's ability to transport these elements through ourselves in the blood stream. At the same time, water plays an important role in transporting waste material out of our bodies.

2.2 An astronomical coincidence?

Ever since we discovered that there was something outside our own world, *the mysterious outer space*, people have wondered if there could be life elsewhere in the universe. Not too long ago, people were still talking about moon creatures and Martians. However, as we explore more and more of our own solar system, we find no present life at all, leaving us with the question, "What made the earth so special?" Why did life evolve exactly here and not anywhere else in our solar system? The answer is water.

Somehow during the creation of the earth, a large amount of water ended up on the face of our planet. So, how did the water form in the first place? And why is it still here after so many millions of years? The simplest answer, and probably the one that comes closest to the truth (at least if we stick to the present scientific knowledge and eliminate any kind of religious beliefs) can be described as a considerable amount of good luck! In more scientific terms, recent theories may be described as follows.

During the early stages in planet formation, gravitational forces created first preliminary bodies, which later collided, forming the basis of the earth. At first, the collision of the larger bodies released an enormous amount of heat, causing the larger elements, such as iron to migrate to the centre. At this stage, lighter elements, such as water, were still in orbit around the sun, but as the earth grew bigger these lighter elements were also pulled in by gravitation and trapped within the earth's crust.

At some point after the sun had ignited and the solar system roughly stabilised, the gases trapped within the crust of the earth were released, forming some kind of atmosphere, mainly containing carbon dioxide, ammonia and methane. The temperature of the earth at this point had dropped to levels where some of the gases condensed to form the oceans we know today.

Did you know...?

- The most common substance found on earth is water. Water is the only substance found naturally in three forms: solid, liquid, and gas.
- Water serves as nature's thermometer, helping to regulate the earth's temperature.

Since the early formation of the oceans an estimated 0.2% of the total amount of water has been lost to space. However, processes similar to the ones that originally created the oceans continuously replace the lost water. The water you see on the earth today is thus the very same water that degassed from the crust after the earth was formed and, for example, the very same water that the dinosaurs drank.

The reason that the water is still here might have to do with the evolution of life. The first photosynthetic life-form produced free oxygen, which replaced the methane and ammonia within the air, creating an atmosphere capable of regulating the temperature of the earth. If the algae had not formed, the earth would not have been able to withstand the increasing solar radiation and the oceans would thus have evaporated.

Once again we have been incredibly lucky that the final temperature of the earth ended up somewhere between the freezing and boiling point of water so that the water does not completely evaporate, nor does the entire earth freeze over.

2.3 Is there enough?

Since the total amount of water on earth does not change much, you might think, "of course there is enough." But is there enough water for us, the human race? How much of the Earth's water is really available for our use?

About 97% of all water on earth lies in the oceans as salt-water and is thus not available for human consumption; at least not without applying enormous amounts of energy to make it into fresh water. Of the remaining 3% of fresh water, the majority is stored in ice caps and glaciers. Only about 0.6% can actually be utilised as long as it is within reach and not contaminated. Table 2.1 shows the distribution of water on earth.

Did you know ...?

- At the present time, about 70% of the surface of the earth is covered with water. The present coastlines are where they are because some of the water is locked up in the polar ice caps.
- If the water locked up in polar ice were to completely melt, the oceans would rise about 75 meters above their present level.

Compared to the roughly 6 billion human beings on the earth today, along with all the other species relying on fresh uncontaminated water, the world's source of consumable water might not continuously be sufficient. As you read this, the world population is exponentially increasing, as is the demand for water. Already many countries are considered to be water-scarce and many are struggling to find new sources of uncontaminated drinkable water.

Table 2.1 Distribution of the world's water resources

Water source	Water volume, in km^3	Percent of total water
Oceans	1,320,000,000	97.24%
Icecaps, glaciers	29,000,000	2.14%
Ground water	8,330,000	0.61%
Fresh-water lakes	125,000	0.009%
Inland seas	104,000	0.008%
Soil moisture	67,000	0.005%
Atmosphere	13,000	0.001%
Rivers	1,250	0.0001%
Total water volume	1,358,000,000	100%

Did you know...?

- It takes 50 litres of water to process 1 litre of beer.
- It takes 150 m^3 of water to manufacture a new car, including tyres.
- 10 litres of water are needed to refine 1 litre of gasoline.
- About 7 m^3 of water are needed to produce the cotton in a pair of jeans and 1.5 m^3 to produce the cotton in a shirt.
- It takes about 4 m^3 of water to grow the wheat to make a loaf of bread.
- Producing an average-size Sunday newspaper requires about 600 litres of water.
- Every glass of water brought to your table in a restaurant requires another two glasses of water to wash and rinse the glass.

The earth is currently undergoing a dramatic change. Within the past century, the human race has developed a capacity to modify earth's basic life support systems and alter the global geophysical processes, which took millions of years to evolve, in matter of decades. We are facing serious consequences for our current and future generations in the way that we are contaminating our freshwater resource. The demand for water is continuously increasing and at the same time the amount of usable water is quickly decreasing.

The scientific community has acknowledged that human interference in the global climate is now evident and that the hydrologic cycle (see below) will be seriously affected in ways we are only now beginning to study.

Did you know ...?

- Half of the world's people lack basic sanitation services. More than a billion lack potable drinking water. In much of the world these numbers are rising, not falling.
- The amount of land irrigated per person is falling, and competition for agricultural water from cities is growing.
- Political and military conflicts over shared water resources are on the rise in some regions.
- Groundwater overdraft is accelerating. Unsustainable groundwater use occurs on every continent except Antarctica.

2.4 Historical perspectives

Humans have generally settled near convenient sources of water. Most of the great ancient civilisations depended on a particular source of water. For example, the Egyptians centred their civilisation on the Nile. Mesopotamia (Greek for the land between the rivers, the Tigris and the Euphrates) was the home of several important ancient empires. Chinese civilisation was located principally in the Yellow and Yangzi river basins. These great civilisations of course considered water as a fundamental aspect.

Water facilitated relatively rapid transportation up until around 1850. In the era of exploration and discovery from the late 15th through to the 18th century, Europeans explored all major oceans and seas.

Water was also an important source of power in the period before the Industrial Revolution. Even though steam power made waterpower less necessary, water remained an essential component in all kinds of manufacturing processes. Beginning with the Industrial Revolution, however, water increasingly becomes a hidden factor in human history. For many, it quite literally went underground, hidden from sight until one turned on a faucet or flushed a toilet.

Water has played an important role in most religions as the source of life. As an example, let us look at the Western European religions.

The first mention of water for Judaism, and later Christianity, is found in the first verse of Genesis: "In the beginning, God created the heavens and the earth. The earth was without form and void, and darkness was upon the face of the deep; and the Spirit of God was moving over the face of the waters" (Gen. 1:1-2).

After creating the heavens and the earth, God then creates light and separates it from the darkness: "And God said, 'Let there be a firmament in the midst of the waters, and let it separate the waters from the waters"' (Gen. 1:6). Order is now imposed on water. "Cosmos" has been created out of "chaos".

Opposite the role of water as the giver of life, water is also, in most religions, considered to be dangerous. It destroys, drowns, and kills by its mere absence. Water cannot be formed, tamed, or shaped by human hands. Only the divine has the capacity to command water. Humans are at the mercy of water, much like they are at the mercy of the divine.

Did you know ...?

- In Assyro-Babylonian mythology, first the gods and subsequently all beings arose from the fusion of salt water (Tiamat) and sweet water (Apsu).
- The holy books of the Hindus explain that all the inhabitants of the earth emerged from the primordial sea.
- In the Koran are the words "We have created every living thing from water".
- In Judaeo-Christian culture, God is called "the fountain of living waters" (Jeremiah 2.13).
- In China, water is considered the specific abode of the dragon, because all life comes from the waters.
- Aphrodite, the ancient Greek goddess of love, was born of the sea.
- In Christianity, baptism links the concepts of the water of life with the waters of purification.
- In India, the sacred River Ganges embodies for Hindus the water of life.
- In Japan, water prefigures the purity and pliant simplicity of life.
- In the cosmogony of Mesopotamian peoples, the abyss of water was regarded as a symbol of the unfathomable, impersonal Wisdom.
- The Roman philosopher Seneca declared that "Where a spring rises or a water flows there ought we to build altars and offer sacrifices".

2.5 The water cycle

Looking at the earth from outside it behaves as a "closed system," meaning that the Earth neither gains nor loses much matter, including water. The same water is continually moving around, through, and above the Earth as water vapour, liquid water, and ice, continually changing its form.

To envision the water cycle, let us take an imaginary journey, following the water through one loop. Because we are dealing with a closed loop we could start practically anywhere water is present. However, since most of the water lies within the pool of the oceans, let us begin here.

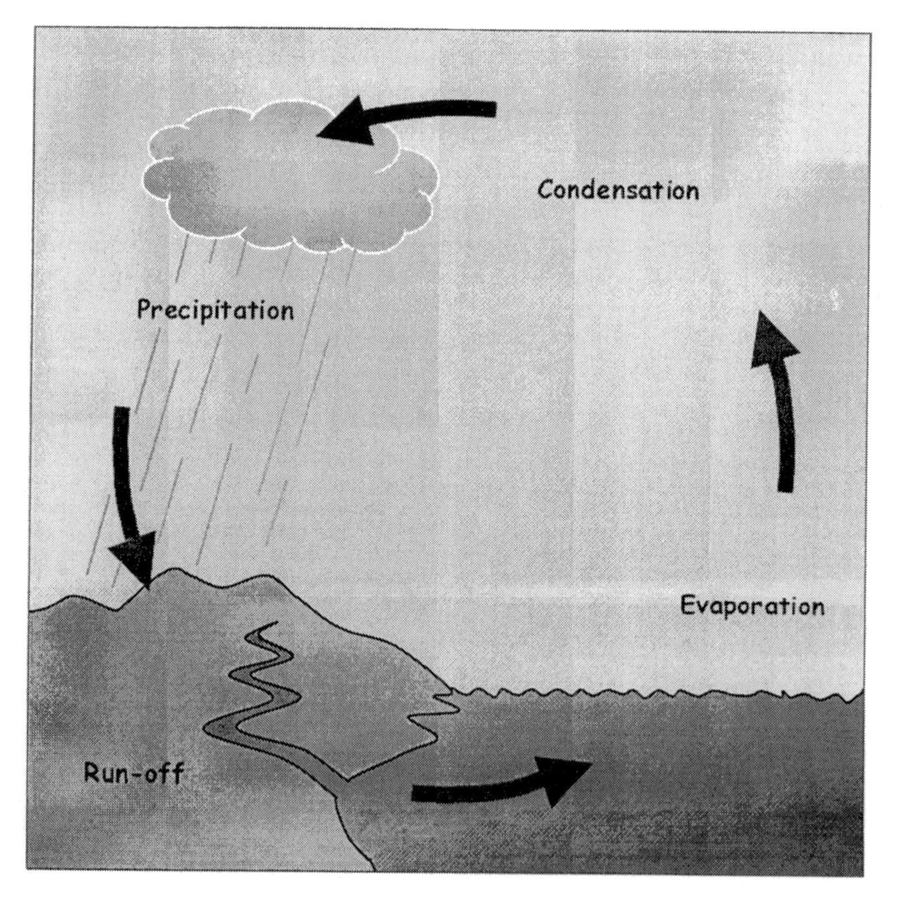

Figure 2.1 Simplified illustration of the water cycle

When sunlight hits the face of the ocean, some of the radiation is absorbed as heat, warming the uppermost layers of water and consequently causing it to evaporate as water vapour. This does not only happen in the oceans, but anywhere sunlight hits surface water, wet soil, or vegetation.

The evaporated water vapour rises into the air. It continues to rise as long as the surrounding air is colder. Then, the water vapour is carried around by the winds until it reaches land. Once over land the vapour is lifted by the updrafts coming from the heated ground surface. Continuously rising, the vapour reaches high in the air where the temperature is quite low. The cold air cools the vapour so it condenses (turns into liquid water again). The condensation builds around tiny dust particles in the air. It is these small droplets we see as clouds in the sky. The droplets collide due to the attractive forces the water molecules in-between (as explained later) growing larger and larger. At some point they can no longer stay suspended in the air and so fall towards the earth as precipitating rain. Depending on the temperature, the condensation may instead create ice crystals falling as snow.

Precipitating either as rain, snow or hail, the water may land anywhere. Maybe the water would land on a leaf in a tree, in which case it would probably evaporate and begin its process of heading for the clouds again.

Did you know...?

- Every 24 hours about 1,000 km^3 of water evaporates from the sea and the land.
- Once evaporated, a water molecule spends an average of 10 days in the air.
- 40% of the atmosphere's moisture falls as precipitation each day.
- About 13,000 km^3 of water, mostly in the form of water vapour, is in the atmosphere at any time. If it all fell as precipitation at once, the Earth would be covered with only about 1 inch of water.
- The Hydrologic cycle in one day uses more energy than humankind has created throughout its entire history.

The water could also land on a patch of dry dirt in a flat field. In this case it might sink into the ground to begin its journey down into an underground aquifer as ground water. The water will continue moving as ground water, but the journey might end up taking tens of thousands of years until it finally gets out of the ground again. However, the water could be pumped out of the ground and be sprayed on crops (from where it might evaporate, flow along the ground into a stream and further into the ocean, or seep back down into the ground). The water might be pumped up from a well, ending up in your faucet or being used to wash a car or clean the floor in some dairy industry. From there, the water returns to the cycle, either by evaporating or by running down the drain, through the sewers and eventually into the ocean.

Instead of soaking into the ground, the water might take a short-cut by landing on your rooftop, a driveway or a car park, in which case it will go down the drain, and into the ocean again.

If no-one interferes, the journey back to the ocean will be relatively fast. However, with 6 billion people needing water for almost anything, there is a good chance that the water will get picked up and used before it gets back to the sea. A lot of surface water is used for irrigation. Even more is used by power-production facilities to cool their electrical equipment. The water might go into the cooling tower to be evaporated. The possibilities are endless, but eventually the water will get back into the environment. From there it will again continue its cycle into and then out of the clouds.

We interfere with the water cycle by taking away huge amounts of freshwater and depleting other water supplies. By clearing vegetation from land to build roads, car parks, etc, water cannot seep into the ground to be stored in the aquifer. As a result of this, the water remains on the surface and increases the likelihood of flash-floods and surface run-off.

2.6 What about water quality?

Just because you have a well that yields plenty of water doesn't mean you can go ahead and just take a drink. Because water is such an excellent solvent it can contain lots of dissolved substances, and since ground water moves through rocks and subsurface soil, it has a high opportunity to dissolve substances as it moves. For that reason, ground water will often have more dissolved substances than surface water will.

Even though the ground is an excellent mechanism for filtering out particulate matter, such as leaves, soil, and bugs, dissolved chemicals and gases can still occur in large enough concentrations in ground water to cause problems. Underground water can get contaminated from industrial, domestic, and agricultural chemicals from the surface.

Contamination of ground water by road salt is of major concern in some areas. Salt is spread on roads to melt ice, and, with salt being so soluble in water, excess sodium and chloride is easily transported into the subsurface ground water.

The most common water-quality problem in rural water supplies is bacterial contamination from septic tanks, which are often used in rural areas that don't have a sewage-treatment system. Wastewater from a septic tank can seep down to the water table and maybe into a home owner's own well. Other self-indulged problems threatening to destroy our water quality include spills of oil and chemicals from industrial activities, use of pesticides and insecticides, leakage of nuclear waste containers, aggressive farming, etc. Actually, there exist very few non-human caused threats to our global fresh water supply.

2.7 So, what is water?

Water plays an important role as a chemical substance. Its many important functions include being a good solvent for dissolving many solids, serving as an excellent coolant both mechanically and biologically, and acting as a reactant in many chemical reactions. Blood, sweat and tears... all solutions of water.

What are the physical and chemical properties of water that make it so unique and necessary for living things? When you look at water, taste and smell it - well, what could be more boring? Pure water is virtually colourless and has no taste or smell. But the hidden qualities of water make it quite an interesting subject.

Did you know...?

- Ice is actually less dense than water, which is why icebergs float on water.
- Water has the highest surface tension of all liquids. Because water can form a "skin" precipitation is possible.
- Water is the greatest reactant the world has ever known.
- There is no force (in existence) that can compress water, meaning nothing could press water into an ice cube.
- Water can freely pass through our cell membrane.
- Water is the "universal solvent". It dissolves more substances than any other liquid.
- Pure water has a neutral pH of about 7, which is neither acidic nor alkaline.

Water has a very high surface tension. In other words, water is sticky and elastic, and tends to clump together in drops rather than spread out in a thin film. Surface tension is responsible for capillary action, which allows water (and its dissolved substances) to move through the roots of plants and through the tiny blood vessels in our bodies.

Water is called the "universal solvent" because it dissolves more substances than any other liquid. This means that wherever water goes, either through the ground or through our bodies, it takes along valuable chemicals, minerals, and nutrients.

Water has a high specific heat index. This means that water can absorb a lot of heat before it begins to get hot. This is why water is valuable to industries and in your car's radiator as a coolant. The high specific heat index of water also helps to regulate the rate at which air changes temperature, which is why the temperature change between seasons is gradual rather than sudden, especially near the oceans.

As chemists we consider water from many perspectives. It is our role to use physical and mathematical laws in application for useful purposes, including diverse perspectives such as living systems, materials and energy. The world of the chemist is a small world - atomic, molecular - which plays a large part in making our lives healthy, comfortable, and hopeful. Due to the diversity of the chemical world, it would be difficult to touch upon all of the applications of water. For the same reason, it would be impossible to discuss the chemical aspects of water without touching upon the physical, mathematical, and biological aspects of the subject.

Did you know...?

- At sea-level pure water freezes into ice at 0°C.
- At sea-level pure water boils into steam at 100°C.
- Seawater freezes at about –2°C.
- Seawater is usually about 3½% heavier than fresh water because it contains about 35 grams of salts per litre.
- Water expands by nearly one tenth of its volume when it freezes. 1 m^3 of water becomes 1.09 m^3 of ice.
- When 1 m^3 of water at sea-level pressure boils away, it becomes about 1,700 m^3 of steam.

Just as you might say that the shape of a key determines its function - which doors it can and cannot open - the structure of a molecule and its composition absolutely determines its functions and properties.

2.8 Water is weird!

Chemically speaking, water is really not common at all. When compared to other compounds of similar molecular size and structure, it is absolutely unique! In fact, its properties are so unusual that it would be irreplaceable. Let us take a chemical look at these unusual properties.

Usually, the solid state of a compound is much denser than the liquid state and thus sinks. When a solid is formed, the molecules become more tightly packed together. When things melt, the molecules move apart. But it is different with water; the solid state of water is less dense than the liquid.

Did you know that the Celsius temperature scale was based on the two physical changes of water? Not because water has typical chemical behaviour, but because water is a familiar substance. However, the boiling point of water is not always 100°C, but depends on the pressure of the air around it. Compared to other chemical substances, water is way out of line! It boils at an extremely high temperature for its size.

Why does water then behave so strangely? The answer lies basically in the geometry of the water molecule and the very special bonds created between water molecules. The bonds - and there are a lot of them - want to hold the water molecules together. Thus, in their liquid state, the water molecules are packed as closely as possible.

When water freezes, the molecules have to arrange themselves in a more systematic way, which is the reason why ice has a lower density than liquid water. Due to the intermolecular bonds, the transformation from liquid state to solid state is relatively easy, which is why the freezing point of water is much higher than expected from the molecule size. The process of boiling requires that the molecules come apart. Due to the strong bonds between the individual water molecules, it takes a lot of energy to do so, which is why the boiling point of water is so high compared to its size. Another result of the water-bonding network is that water has a very high specific heat. Once heated, water takes a very long time to cool off. Or conversely, it takes a lot of heat to make water hot.

We are actually quite lucky that water is weird. If water were to behave in a "normal" way compared to its molecular size, it would be a gas at room temperature, so thank the strong bonds between water molecules for oceans, rain, and body fluids.

3 Physical properties of water

Erik Trostmann
Technical University of Denmark

3.1 Water as a pressure medium

When a shear force, no matter how small it may be, is applied to a fluid (whether it is a gas or a liquid) the fluid is characterized by being under continuous deformation. Liquids and gases are distinguished by their respective compressibilities. In fluid power control the term "hydraulic" is used to characterize a system using a liquid as pressure medium and "pneumatic" is used to characterize a system using a gas as pressure medium.

Many different types of liquids are being used for hydraulic control and power transmission systems. They may range from water, to mineral oils, to synthetic and organic liquids, to vegetable oils or to molten metals. Water was the first hydraulic pressure medium to be used in the early days. Water is available everywhere, it is cheap and rather efficient for power transmission. When mineral oil was discovered in the beginning of the 20th Century it soon became the preferred pressure medium in hydraulic systems, due to its many superior properties.

During the last decades of the 20th Century a new focus has been directed towards water - pure tap water - as a pressure medium for hydraulic systems. The increasing demands from environmental concerns have created a growing mistrust in using mineral oil in hydraulics. Industry must pay an increasing attention to:

- improved safety at the working place
- improved environmental protection

These requirements have added new criteria for the selection of pressure media for hydraulic systems such as:

- non-flammability
- product compatibility
- biological degradability

Good non-flammability means that the hydraulic medium does not pose any fire risk. Good product compatibility means that the produced product will not be contaminated by leaks from the hydraulic fluid when a hydraulic-powered production system is used. Good biological degradability means that no long-lasting contamination of production systems, floor, soil or ground water etc will occur if the hydraulic system leaks.

Taking the above new criteria as well as the traditional technical criteria for viscosity, lubrication and wear, compressibility and aging properties into consideration makes water and especially clear water (pure tap water) an interesting and attractive medium for hydraulics. However, the physical properties of water differ significantly from those of mineral oil.

A change over of the pressure medium from mineral oil to water can, in general, not be done simply by modifying the oil components assuming that the performance and lifetime should be maintained. The significant difference in physical properties of the two types of pressure media, water and mineral oil, requires a shift in design paradigm.

Such a shift has been carried out by the Danish company Danfoss A/S in 1994 after more than 5 years of research and development efforts. In their new design paradigm Danfoss has taken advantage of (1) new ingenious designs (2) applying new materials, such as anodized aluminum, stainless steel, bronze and polymers and (3) using hygienic design concepts.

Since a hydraulic control system requires a physical well defined pressure medium for its operation, considerably detailed knowledge of the properties of such a medium is important.

The purpose of this chapter is to define and review the most important physical properties of water pertaining to the application of pure tap water as a pressure medium for water hydraulic control systems.

Although several quantitative data and graphs will be illustrated the presentation cannot be exhaustive. The focus is directed towards the temperature range 0-100°C and the pressure range 1-300 bar and comprehends mainly fluid water's bulk properties, i.e. thermodynamic, transport and electric properties.

The most significant differences in physical properties between water and mineral oil are summarised and compared in the following sections.

3.1.1 Molecular structure of liquid water

Water is a unique substance with properties that are quite anomalous to other substances. It is closely associated with cleanliness, freshness and purity. The existence of water in nature in all three phases (i.e. as ice, liquid water and water vapour), and its profusion makes this planet unique.

Of all known liquids water is probably the most studied and the least understood. In spite of this, many of its physical properties such as density, mass and viscosity are accepted as international standards at its triple point, at which ice, liquid water and water vapour co-exist at equilibrium. The triple point for water is at 0.01°C and 611 Pa (4.58 mm Hg) [17].

In comparison to other solvents water shows unique properties that are highly sensitive to temperature and pressure changes. These properties stem from its structure and the electrical charge distribution of the water molecule.

The exact molecular structure is not known. Many conceptual rather than physical models have been proposed. A popular model theory is the so-called

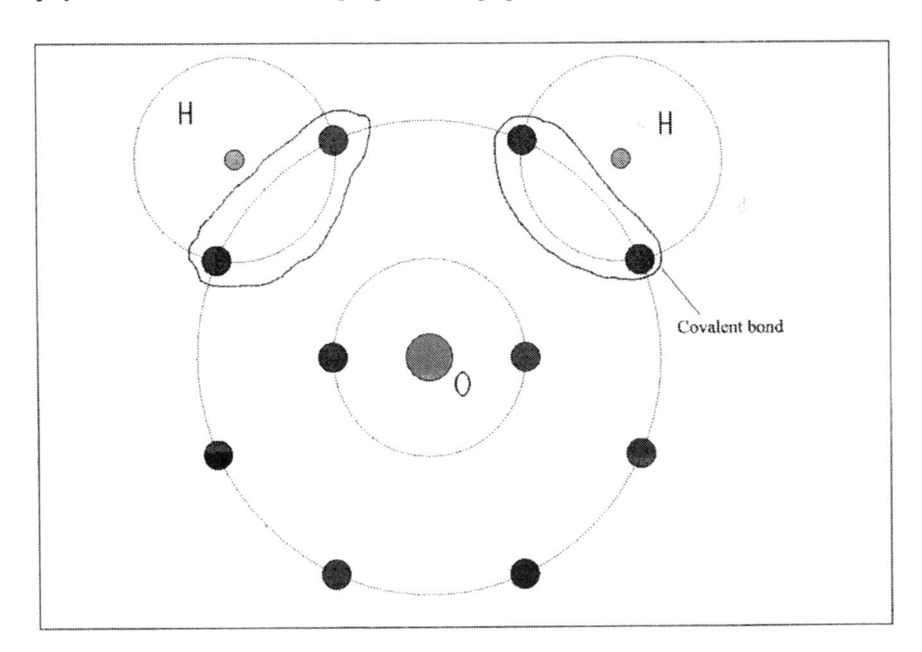

Figure 3.1 A schematic, anatomical view of a water molecule

free-volume theory, where the excess volume of the water molecules is considered as holes, known as vacancies [2]. These vacancies qualitatively account for the mobility, density and entropy of liquids.

Accordingly, the water molecule consists of a combination of two hydrogen atoms with one electron in their outer (and only) shell and one oxygen atom with two electron shells containing 6 electrons in its outer shell and 2 electrons in its inner shell.

The hydrogen atoms can each hold two electrons in their shell and therefore each one has one vacancy. These vacancies can be filled by sharing two outer shell electrons from an oxygen atom. In this way a water molecule is formed, see Figure 3.1.

The above mentioned sharing of electrons is known as a covalent bond. The two covalent bonds are indicated in Figure 3.1. Covalent bonds are very strong because much energy is needed to break them down.

The resultant molecule of H_2O is called a polar molecule since its positive and negative charges (δ + and δ -) are not distributed evenly around a centre but are placed asymmetrically; therefore positive and negative poles are formed.

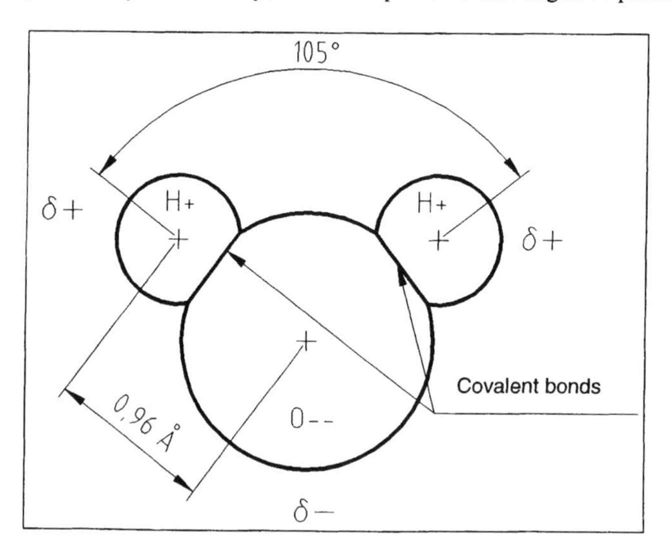

Figure 3.2 A schematic external view of a water molecule. The covalent bonds are indicated as well as the electrical charge (δ+ and δ–)

In Figure 3.2 an over simplified schematic illustrates how two hydrogen atoms are attached to one oxygen atom, to form a water molecule. The equilibrium position occupied by the two hydrogen atoms positions them 0.96 Å from the oxygen nucleus. The angle of separation of the two positive charges of the water molecule is approx 105°[3].

A chemist would describe this schematic process by writing: $H_2 + O = H_2O$. The H_2O formula should actually be H_4O_2 because two H_2 molecules and one O_2 molecule participate in the reaction. For the sake of simplicity it is assumed that each molecule H_4O_2 breaks apart immediately and forms two molecules H_2O.

3.1.2 Bonds between water molecules

The asymmetry of the water molecule leads to a positively charged side where the hydrogen atoms are bonded, and a negatively charged side to the opposite. The charges, δ + and δ -, act like the poles of a magnet and generate an attractive force between water molecules, see Figure 3.3, which is called a hydrogen bond.

Hydrogen bonds are weak compared to the strength of covalent bonds, but they are much stronger than the intermolecular bonds present in liquids with symmetrical nonpolar molecules [3].

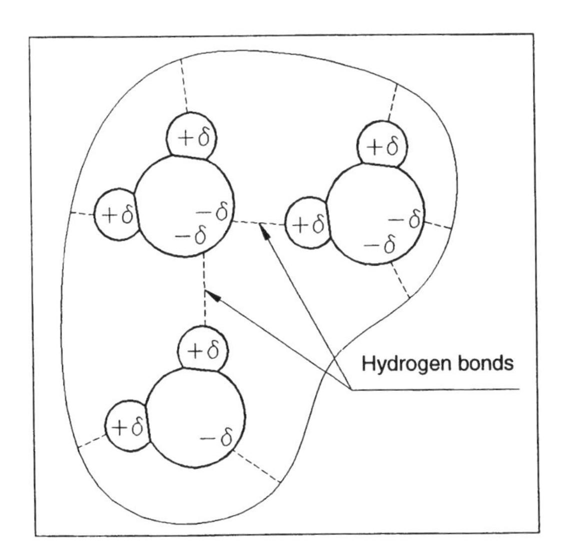

Figure 3.3 A model of hydrogen bonding between water molecules. The dashed lines between molecules represent hydrogen bonds

Each water molecule can be involved in up to a maximum of four hydrogen bonds, see Figure 3.3. Water molecules, through their hydrogen bonds, show cohesion. One effect of cohesion is surface tension, which allows water droplets to bead up.

Liquid water has a partially ordered structure in which hydrogen bonds are constantly formed and broken up. When water exists in its liquid form, most hydrogen bonds are present, but not every molecule is committed to the maximum bonds. In liquid water each molecule is hydrogen bonded to approx 3.4 other water molecules.

3.1.3 Science versus empiricism

In spite of all the effort devoted over the last century to the study of water, we are still far from a full understanding of the exact nature of the molecular interactions in the liquid [2].

Classical thermodynamics does not reflect direct information about the microscopic behaviour of liquids. But if bulk properties are applied in relation with a simple molecular theory of matter some useful information can be found.

The simplest comparison to be made is between thermodynamic and transport properties of materials under similar conditions. A striking feature is the similarity between the liquid and solid states for some relevant properties of, for instance, argon, benzene, water and sodium: small increase in density, small differences between latent heats of fusion and evaporation and similarity of specific heats. This indicates that those liquids are highly condensed with strong intermolecular cohesive forces and suggests that the liquids and solids are similar in the arrangement of their molecules. Although such gross observations are being made it is still data of rather qualitative nature.

A quantitative approach for studying the microscopic behaviour of liquids has been proposed by using equations of state like the equation of state for gases. Many attempts have been made, but no general formulation of such equations has so far been successful.

The bulk properties of liquids arise from the average of a large number of molecular interactions. In theory each molecule's motion could be considered separately and likewise all attractive and repulsive forces influencing the molecules should be observed. Then a summation of these effects across time and all molecules should be made. It should then be theoretically possible to predict any of the bulk properties. However, due to the complexity of the problem no general solution has been found so far.

The bulk properties of water relevant for applying water as a pressure medium in water hydraulic control systems will, for the time being as well as for the nearest future, continue to rely on empiricism, experimental data and experience.

3.2 Properties of liquid water

3.2.1 Thermodynamic properties

a. Density

Mass density is the mass (or quantity of matter) contained in a unit volume of the substance under consideration. The mass density is denoted by the symbol ρ and derived as the ratio of a mass M over its volume V:

$$\rho = \frac{M}{V} \tag{3.1}$$

The mass is independent of the pressure and temperature. Changes in the pressure and temperature state in a hydraulic pressure medium will change the value of its density and of the hydraulic energy losses in the system, because the latter is proportional to the density.

The density of the pressure medium in a hydraulic system ought to be low, in order to keep the pressure losses small and to reduce dynamic feedback effects on control valves.

The weight density γ is defined as the weight F per unit of volume V:

$$\gamma = \frac{F}{V} = \rho \cdot g \tag{3.2}$$

where g is the acceleration of gravity, $g = 9{,}81 \mathrm{m/sec}^2$.

The specific volume ν, which is the reciprocal of the specific weight, is often used in the discussion of the behaviour of gasses.

The relation between weight density and mass density is given by:

$$\rho = \frac{\gamma}{g} \tag{3.3}$$

Specific gravity is the ratio of the mass (or weight) density of a substance at a certain temperature to the mass (or weight) density of water at the same temperature. Specific gravity is dimensionless and the symbols σ or SG are often used. The temperature, however, must be specified.

In the USA the petroleum industry has selected 60°F as a standard temperature for specific gravity of hydraulic pressure media. The specific gravity of an oil at 60°F over the specific gravity of water is often named σ60/60°F [4].

In Europe the density of hydraulic media is often referred to the temperature 15°C (~ 59.0°F) and to the pressure 1 bar (abs).

The density of most hydraulic media decreases with increasing temperature and increases with increasing pressure. Water is an exception [12]. Its density

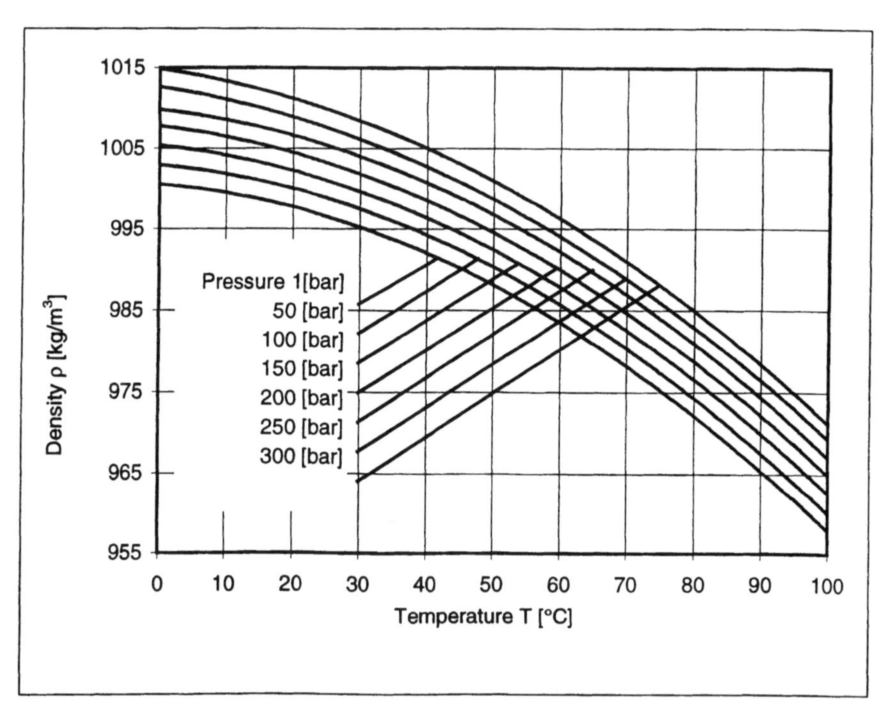

Figure 3.4 Mass density, ρ of pure water as a function of temperature, T and pressure, P

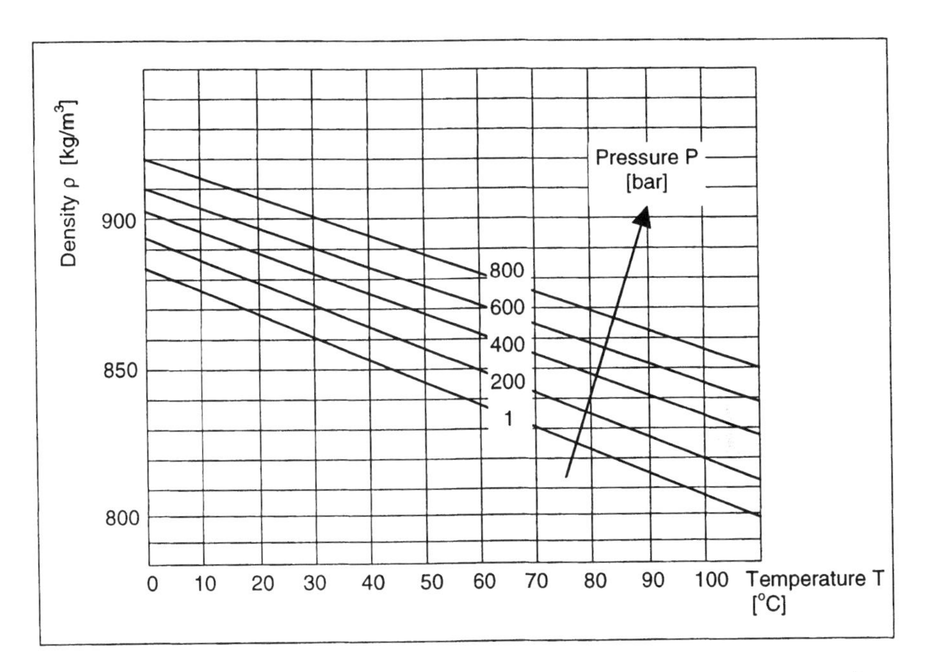

Figure 3.5 Mass density, ρ of a mineral oil of a density, ρ = 864 kg/m³ at reference temperature, T = 15°C

increases to a maximum when temperature increases from 0°C to its maximum at 3.98°C and then it decreases for increasing temperature above 3.98°C.

In Figure 3.4 the density of pure water is illustrated as a function of temperature T in the range 0-100°C, and of pressure P in the range 1-300 bar (abs pressure).

In Figure 3.5 the density of a mineral oil, as a function of temperature, T and pressure, P, is shown for the range 0-100°C and 1-800 bar (abs pressure) and reference pressure 1 bar (abs pressure) as a function of temperature and pressure [5].

In Figure 3.6 a comparison of the amount of compressibility of a mineral oil and water is illustrated [7]. It is noted that the variation of the relative mass densities as a function of pressure shows an increase somewhat higher for mineral oil than for water.

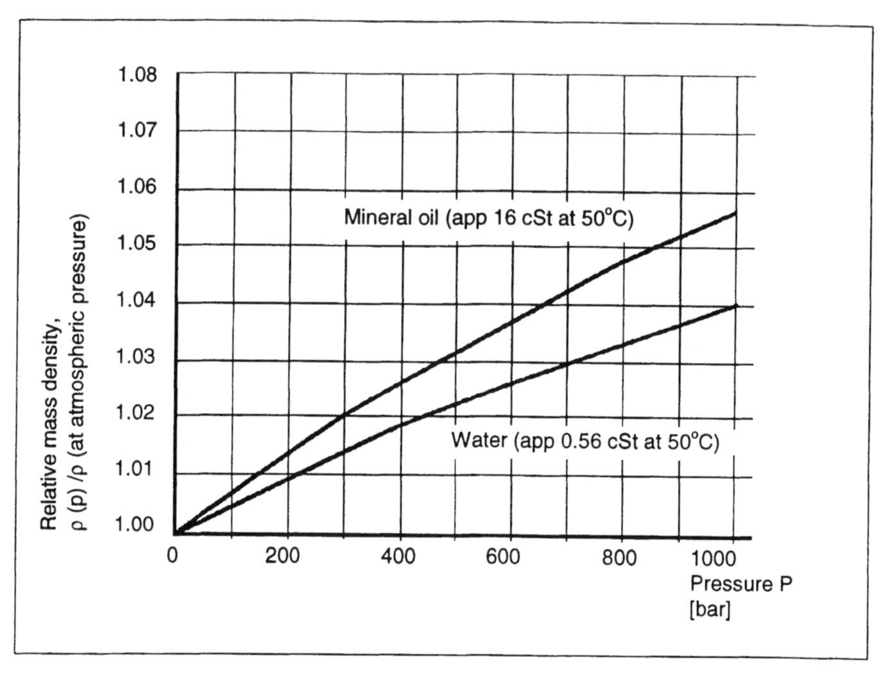

Figure 3.6 Relative mass density, ρ(P)/ρ (at atmospheric pressure ~ 1 bar) as a function of pressure

The density ρ(T) as a function of temperature T can be approximated by the empirical equation (3.4), [9]:

$$\rho(T) = \rho(@15°C)\cdot[1-\alpha_P(T)\cdot[T-15°C]] \tag{3.4}$$

where the cubic volume expansion coefficient (isopiestic coefficient of thermal expansion) $\alpha_P(T, P)$ for pure water is shown in Figure 3.7.

The average isopiestic volume coefficient of thermal expansion, α_P for mineral oils can be estimated to the constant value 0.00075 per °C.

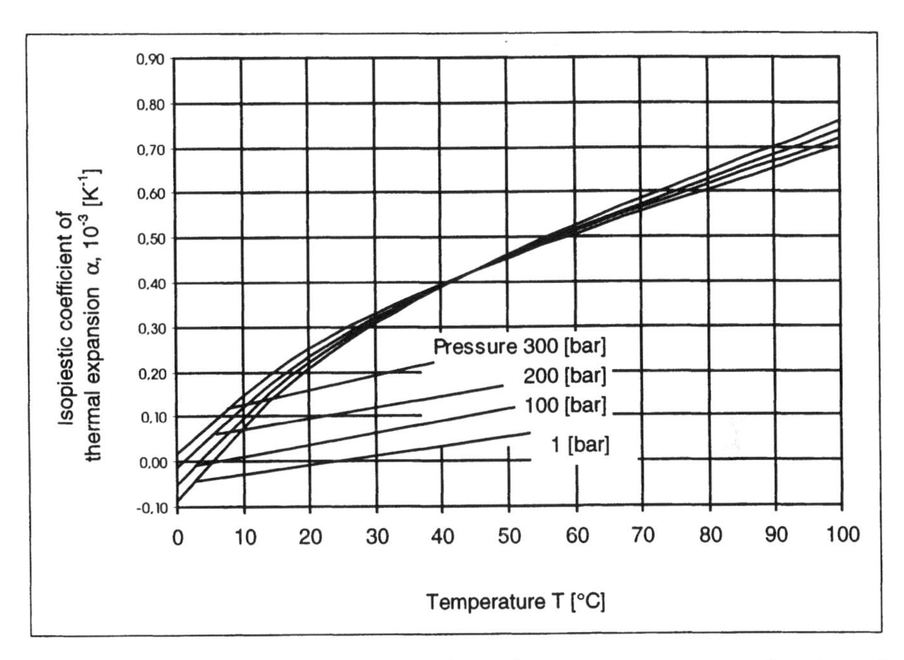

Figure 3.7 Isopiestic coefficient of thermal expansion α_P as a function of temperature and pressure

b. Equation of state

As described above (Figure 3.4 and Figure 3.5) the density of a liquid is a function of both temperature and pressure. A function, which relates density, temperature and pressure in a liquid, is by definition designated as the equation of state for that liquid.

The equation of state for a liquid has not yet been derived mathematically from physical principles or laws, but is empirically based. In contrast to this the equation of state for gases can be derived statistically from the kinetic theory of gases. Fortunately, changes in density caused by changes in temperature and/or pressure are small for a liquïd. Therefore simplified approximate, linear expressions may be sufficient for most practical purposes.

Under certain, usually mathematical, assumptions which are fulfilled here, the state equation for a liquid expressed by:

$$\rho = \rho(P,T) \tag{3.5}$$

can be approximated by a Taylor's series for two variables, using only the first (linear) terms as follows [8]:

$$\rho = \rho_0 + \left[\frac{\delta\rho}{\delta P}\right]_{P_0T_0}(P - P_0) + \left[\frac{\delta\rho}{\delta T}\right]_{P_0T_0}(T - T_0) \tag{3.6}$$

The above equation describes the mass density ρ in a small volume around a reference point (ρ_0, P_0, T_0) in the (ρ, P, T) definition space relevant for the application of the liquid as a function of pressure and temperature.

Compression and decompression of a liquid may be isothermal ("static" or "slow") or isentropic ("dynamic" or "fast") or be a mixture of these thermodynamic processes.

The bulk modulus (of elasticity) of a fluid is introduced as a measure of the fluid's resistance to reduction or expansion in its volume under a varying pressure. In fact two definitions are introduced, an isothermal and an isentropic bulk modulus, β.

The reciprocal quantity of bulk modulus is denoted compressibility, κ. The compressibility of a hydraulic medium determines the amount of energy absorbed during the generation of pressure in a hydraulic system as well as the amount of energy released during a decompression. In this way the compressibility highly influences pressure transients in a system caused by pressure variations.

In most hydraulic systems it is normally specified that the compressibility should be as low as possible. It should, however, be noted that the inherent elasticity of mechanical parts and components in the system tend to increase the apparent total compressibility, the so-called "effective" compressibility above the value of the hydraulic liquid itself. In order to get as low compressibility as possible, the hydraulic liquid should be kept free of undissolved or entrained air, as well as vapour bubbles (see later).

It should also be noted that when the compressibility becomes extremely low, high pressure transients may be more easily transmitted through the pressure medium due to the low inherent damping of hydraulic systems. On the other hand less decompression shocks will occur.

c. Definition of bulk moduli

Based upon the above mentioned equation (3.6), the following generalised definitions [5] are used denoting an isothermal bulk modulus β_T, an isentropic bulk modulus β_S, and also an isopiestic, volumetric expansion coefficient α_P for a liquid:

$$\beta = \rho \cdot \left.\frac{\delta P}{\delta \rho}\right] \qquad \textit{(isothermal bulk modulus)} \qquad (3.7)$$

$$\beta = \rho \cdot \left.\frac{\delta P}{\delta \rho}\right] \qquad \textit{(isentropic bulk modulus)} \qquad (3.8)$$

$$\alpha_P = \frac{1}{\rho} \cdot \left.\frac{\delta \rho}{\delta T}\right]_P \qquad (3.9)$$

where ρ is the density of the liquid at pressure P, and T and S refer to the conditions of constant temperature and entropy, respectively.

In Figure 3.7 the isopiestic coefficient of thermal expansion, α_P is shown as a function of temperature and pressure. It is noted that α_P increases a great deal for increasing temperature but is relative insensitive to pressure variations.

Using the well known relationships:

$$\rho = \frac{M}{V} \quad \textit{and} \quad \frac{\delta \rho}{\delta V} = -\frac{M}{V^2} \qquad (3.10)$$

where V is a volume of the liquid at pressure P and M is its mass, the above definitions (3.7-3.9) can be expressed as [5]:

$$\beta_T = -V \cdot \left.\frac{\delta P}{\delta V}\right]_T = \beta_{T,\ tangent} \qquad (3.11)$$

$$\beta_S = -V \cdot \left.\frac{\delta P}{\delta V}\right]_S = \beta_{S,\ tangent} \tag{3.12}$$

β_T and β_S in the above definitions (3.11) and (3.12) are the exact, scientific bulk moduli used in thermodynamics. These values are frequently termed the tangent moduli $\beta_{T,\ tangent}$ and $\beta_{S,\ tangent}$.

In Figure 3.8 a volume-pressure curve for a liquid at constant temperature is illustrated schematically. The isothermal tangent bulk modulus β_T is measured as the slope of the tangent, the gradient $\delta V/\delta P$ at pressure P and constant temperature [8].

Assuming an isentropic compression of the liquid in Figure 3.8 an isentropic tangent bulk modulus β_S can analogously be defined. Knowing the ratio of the specific heats C_P/C_V the isentropic tangent bulk modulus β_S can be derived by the equation (3.13):

$$\beta_S = \frac{C_P}{C_S} \cdot \beta_T \tag{3.13}$$

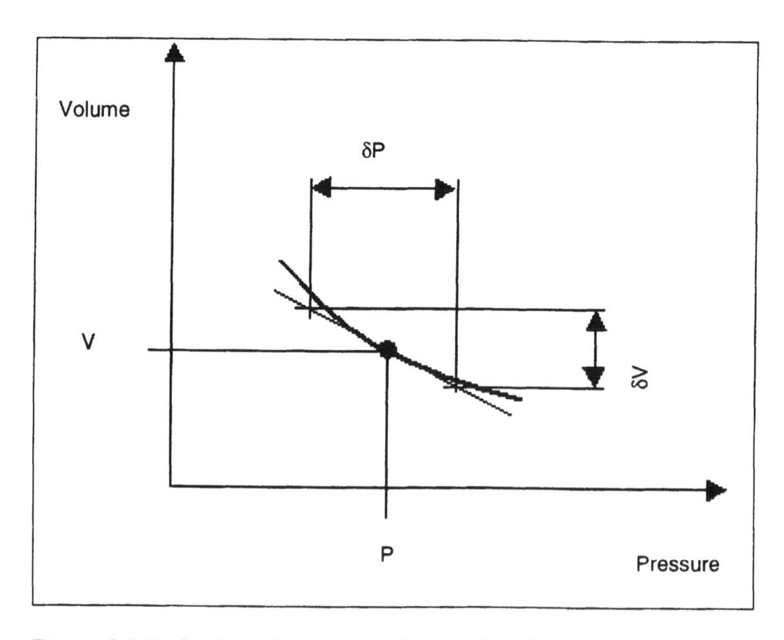

Figure 3.8 Definition of tangent bulk modulus β_T

In engineering practice in particularly USA [5], the following definitions are introduced:

$$B_T = -V_0 \cdot \left.\frac{\delta P}{\delta V}\right]_T$$

$$B_S = -V_0 \cdot \left.\frac{\delta P}{\delta V}\right]_S \qquad (3.14)$$

where V_0 is the volume of the considered fluid at standard pressure (usually the atmospheric pressure, 1 bar (abs)), B_T is the isothermal bulk modulus and B_S the isentropic bulk modulus.

In USA the bulk modulus definitions in (3.14) are considered to be more convenient for practical purposes than the definitions in (3.11) and (3.12) and have therefore been recommended by ASTM as definitions for bulk moduli.

Using ρ and ρ_0 respectively for the mass densities of the volumes V and V_0 the following relations [5] can be derived:

$$B_T = \frac{\rho}{\rho_0} \beta_{T,\ tangent}$$

$$B_S = \frac{\rho}{\rho_0} \beta_{S,\ tangent} \qquad (3.15)$$

Considering a liquid the substitution of V_0 for V at low pressures is not significant. However, at high pressures the deviation becomes large. Considering a gas, the use of V_0 for V can lead to large errors. The above relations, (3.15) may be important in connection to a fluid mixture of a liquid and a gas (such as air or vapour).

Sometimes a third variant of bulk moduli definitions for engineering purposes is applied [5]. In a Figure 3.9 volume-pressure curve for a liquid at constant temperature is illustrated schematically. When a finite volume, V_0 is compressed the volume decreases an amount ΔV. Assuming ΔV is proportional with the increase ΔP in pressure, P an isothermal, secant bulk modulus, $\beta_{T,\ secant}$ is defined by:

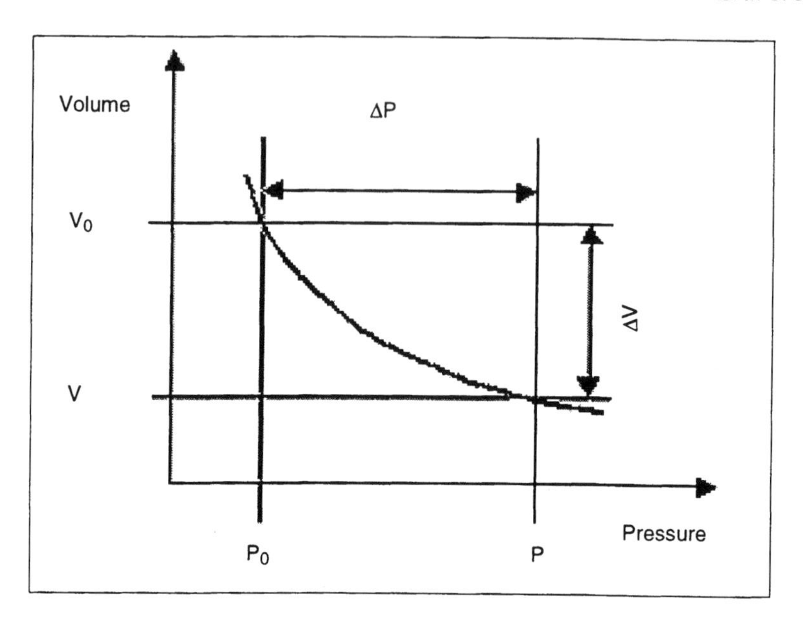

Figure 3.9 Definition of secant bulk modulus β_T [8]

$$\beta_{T,\ secant} = -V_0 \cdot \frac{\Delta P}{\Delta V}\bigg]_T \tag{3.16}$$

Assuming an isentropic compression illustrated for the liquid in Figure 3.9 an isentropic, secant bulk modulus, β_S can analogously be defined by:

$$\beta_{S,\ secant} = -V_0 \cdot \frac{\Delta P}{\Delta V}\bigg]_S \tag{3.17}$$

In the above expressions (3.14), (3.16) and (3.17) V_0 is the initial volume at a reference pressure (~1 bar absolute), ΔV is the difference in volume, V at pressure, P and V_0 at reference pressure. Note that a reduction in volume is by definition, counted negatively, see equation (3.11), (3.12), (3.14), (3.16) and (3.17) the various bulk moduli become positive numbers.

The bulk modulus, β and compressibility, κ vary with temperature and pressure [9]. The isothermal tangent compressibility, κ_T of water is illustrated in Figure 3.10 over the temperature range 0–100°C and the pressure as a parameter in the interval 1-300 bar (abs).

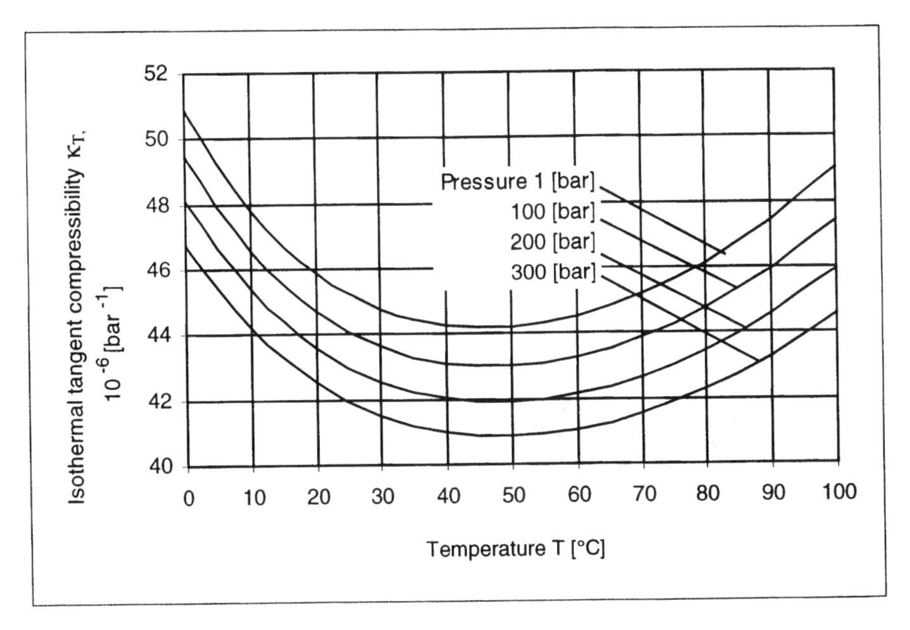

Figure 3.10 Isothermal tangent compressibility κ_T for waterK

Compared to other liquids the compressibility of water is unusually large and possesses a minimum near 46.5°C at reference pressure 1 bar (abs) [9].

From Figure 3.10 this minimum value reads:

$$\kappa_T \ @\ 45.5\,°C \ and \ 1\,bar(abs) = 44.1 \cdot 10^{-6}\,bar^{-1} \left(\beta_T = 22.7 \cdot 10^3\,bar\right)$$

$$\kappa_T \ @\ 50.0\,°C \ and \ 300\,bar(abs) = 40.8 \cdot 10^{-6}\,bar^{-1} \left(\beta_T = 24.5 \cdot 10^3\,bar\right)$$

For comparison the isentropic tangent bulk modulus for a typical mineral, hydraulic oil is illustrated in Figure 3.11 over the temperature range 10–100°C and with pressure as a parameter in the interval 1-700 bar (abs) [5]. For mineral oil bulk modulus (see Figure 3.11) decreases with temperature and increases with pressure. Equivalent values for mineral oil to the above mentioned for water read:

$$\beta_S \ @\ 45\,°C \ and \ 1\,bar(abs) = 15.8 \cdot 10^3\,bar \left(\beta_T = 13.5 \cdot 10^3\,bar\right)$$

$$\beta_S \ @\ 50\,°C \ and \ 300\,bar(abs) = 18.5 \cdot 10^3\,bar \left(\beta_T = 15.8 \cdot 10^3\,bar\right)$$

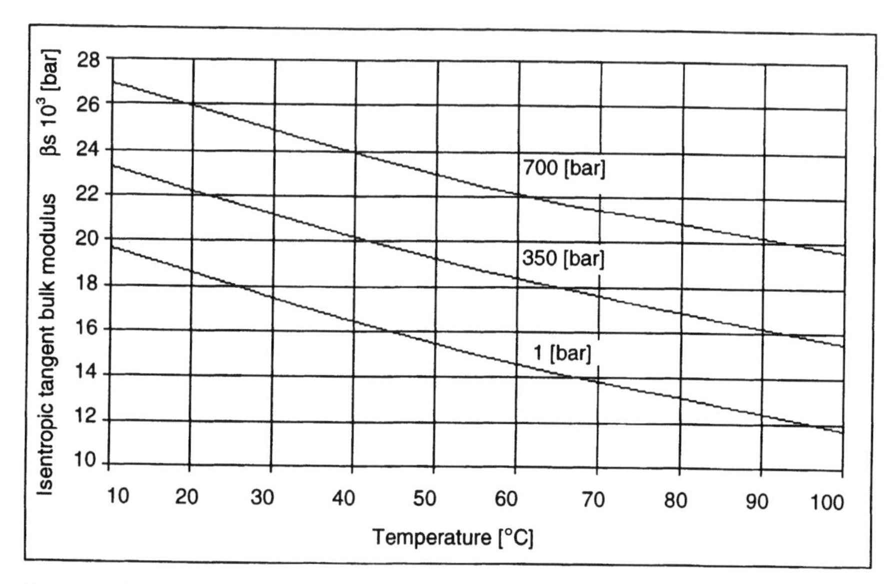

Figure 3.11 Isentropic tangent bulk modulus β_S for a mineral oil

Note that the isothermal and isentropic tangent bulk moduli are related equation (3.13) by:

$$\frac{\beta_S}{\beta_T} = \gamma \qquad (3.18)$$

where γ is the ratio of the specific heats at constant pressure, C_P and constant volume, C_V:

$$\gamma = \frac{C_P}{C_V} \qquad (3.19)$$

The isothermal tangent modulus can be derived from the gradient of the volume-pressure curve shown in Figure 3.8, when the compression process is isothermal and the isentropic tangent bulk modulus can be derived by using (3.18) and (3.19).

The specific heats, C_P and C_V are functions of temperature as well as of pressure. For water typical values are listed in Table 3.1, and for mineral oil typical values are listed in Table 3.2.

Table 3.1 Typical values of C_P/C_V for water [10]

Pressure [bar]	Temperature [°C]		
	10 [°C]	40 [°C]	80 [°C]
1.01325	1.0011	1.026	1.084
200	1.0026	1.028	1.083
700	1.0092	1.035	1.083

Table 3.2 Typical values of C_P/C_V for mineral oil [5]

Pressure [bar]	Temperature [°C]		
	10 [°C]	65.6 [°C]	121 [°C]
1.01325	1.175	1.165	1.155
689.5	1.15	1.14	1.13

Compared with mineral oil the bulk modulus of water is approx 50% greater. This means that the response speed of water is much higher and possible decompression shocks are correspondingly smaller. However, the risk of water hammering is higher when water hydraulic systems are used.

d. Effective bulk modulus

The elasticity of the fluid in a hydraulic system and the elasticity of the walls in the container enclosing the fluid (e.g. motors, cylinders, pipes, fittings, sealings and in particular flexible hoses) add up to the total compressibility of a hydraulic component or system seen from the fluid.

The interaction of the spring effect of the above described elastic members of a hydraulic component or system and their involved masses result in a resonant frequency. Furthermore, due to the inherent poor damping, hydraulic vibrations are introduced, which may drastically limit the dynamic performance of the component or system in question.

The reciprocal value of the above mentioned compressibility is often denoted as the effective bulk modulus, β_e. The effective bulk modulus consists of three contributions: the bulk modulus, β_ℓ of the hydraulic liquid itself, the bulk modulus, β_g of the entrained air/gas and/or vapour and the bulk modulus, β_{co} of the container.

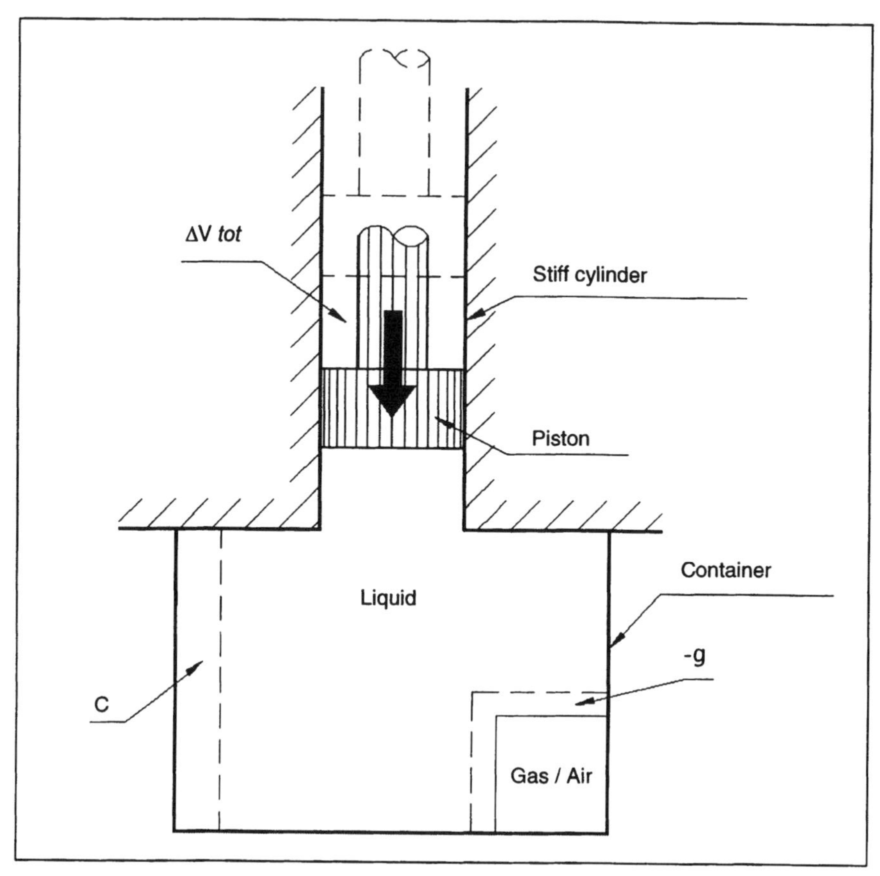

Figure 3.12 Container with elastic walls containing a pressurised mixture of liquid and undissolved gas/air

Assuming a pressure increase, ΔP is exerted on a fluid mixture of a liquid with undissolved gas/air in an elastic container by the motion of a piston, see Figure 3.12 the following relationship for the above-mentioned bulk modulus can be derived [4]:

$$\frac{1}{\beta_e} = \frac{V_g}{V_{tot}} \cdot \frac{1}{\beta_g} + \frac{V_\ell}{V_{tot}} \cdot \frac{1}{\beta_\ell} + \frac{1}{\beta_{co}} \qquad (3.20)$$

where the bulk moduli are defined as follows:

$$\beta_e = V_{tot} \frac{\Delta P}{\Delta V_{tot}}$$

$$\beta_g = -\frac{V_g \cdot \Delta P}{g}$$

$$\beta_{co} = \frac{V_{tot} \Delta P}{c} \tag{3.21}$$

$$\beta_\ell = -\frac{V_\ell \cdot \Delta P}{\ell}$$

and where:
V_{tot} = Initial total volume of the container
V_ℓ = Initial volume of liquid
V_g = Initial volume of undissolved gas/air

and initially the total volume of the container is expressed by:

$$V_{tot} = V_\ell + V_g \tag{3.22}$$

Referring to the motion of the piston,

Figure 3.12, the decrease of the initial volume, V_{tot} can be expressed as:

$$\Delta V_{tot} = -g - \ell + c \tag{3.23}$$

where g, ℓ and c are the volume changes indicated in Figure 3.12 for the initial gas volume, V_g the liquid volume, V_ℓ and the container volume, V_{tot} respectively.

An accurate computation of (3.20) requires knowledge of several quantities. However, the following two simple approximations may be useful.

Assuming the container to be infinitely stiff ($\beta_{co} \approx \infty$) equation (3.20) can be approximated by:

$$\frac{1}{\beta_e} \approx \frac{V_g}{V_{tot}} \cdot \frac{1}{\beta_g} + \frac{V_{tot} - V_g}{V_{tot}} \cdot \frac{1}{\beta_\ell} \approx \frac{V_g}{V_{tot}} \cdot \frac{1}{\beta_g} + \frac{1}{\beta_\ell} \tag{3.24}$$

or assuming no undissolved gas/air (or vapour) to be present, then equation (3.20) reduces to:

$$\frac{1}{\beta_e} \approx \frac{1}{\beta_{co}} + \frac{1}{\beta_\ell} \tag{3.25}$$

In order to use equations (3.24) and (3.25) for computation of β_e the quantities $\beta_\ell, V_g, V_{tot}, \beta_g$ and β_{co} must be known. The bulk modulus, β_ℓ can be obtained from graphs such as Figure 3.10 and Figure 3.11. The volume, V_g of undissolved gas may be estimated. The total volume, V_{tot} can be derived from the system's dimensional data. The isothermal bulk modulus of a gas is equal to the pressure level, P and the isentropic bulk modulus for a gas is $\beta_g = 1.4 \cdot P$. The bulk modulus, β_{co} of the elastic container is difficult to compute or estimate. Some guidance for doing this can be found in references [4] and [11].

As can been seen from equations (3.24) and (3.25) reciprocals are involved so the effective bulk modulus will be less than any one of the values:

$$\beta_\ell, \beta_{co} \text{ and } \frac{V_g}{V_{tot}} \cdot \frac{1}{\beta_g}.$$

The effective bulk modulus, β_e can often be of drastically reduced value compared to the bulk modulus, β_ℓ of the hydraulic pressure medium itself [11]. Especially the use of pressurised flexible hoses and the occurrence of undissolved air in the hydraulic system contribute to this reduction.

In Figure 3.13 and Figure 3.14 estimates of the effect of entrained undissolved air in a hydraulic system is shown in graphical form by using equation (3.24). The ratio β_e/β_ℓ is shown as a function of pressure for various ratios of undissolved air/liquid volumes. Two cases are illustrated: in Figure 3.12 the bulk modulus, $\beta_{\ell-H_2O}$ for pure water is used and for comparison the bulk modulus, $\beta_{\ell\text{-oil}}$ for mineral oil is used in Figure 3.13.

In both cases entrapped, undissolved air reduces the ratio β_e/β_ℓ and especially at low pressures (< 30 bar) the reduction becomes significant. For instance the

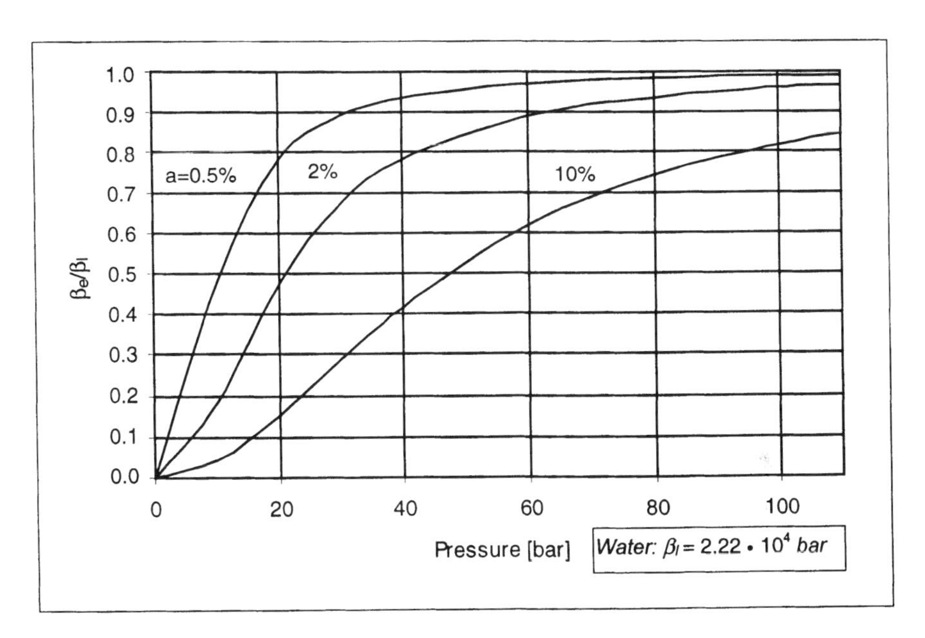

Figure 3.13 Effect of undissolved air/water ratio on bulk modulus

effective bulk modulus is approx halved at 20 bar, and 2% of undissolved air compared to the bulk modulus of pure water $(\beta_{\ell-H_2O})$ or mineral oil $(\beta_{\ell-oil})$.

Note that the relative effect of undissolved air on the effective bulk modulus is greatest, the higher the bulk modulus of the pressure medium itself is. However, for the same volume of undissolved air/liquid mixture the effective bulk modulus in absolute scale is highest for mixtures having the highest bulk modulus.

e. Thermal properties

The specific heat capacity and the thermal conductivity of the pressure medium in a hydraulic system affect the heat balance of the system essentially.

The specific heat capacity at constant pressure, C_P can be defined as the amount of energy (heat) needed to change the temperature of one kilogram of a substance by 1°C. The specific heat capacity at constant pressure for water is high compared to most substances. Exceptions are liquid H, He and Li.

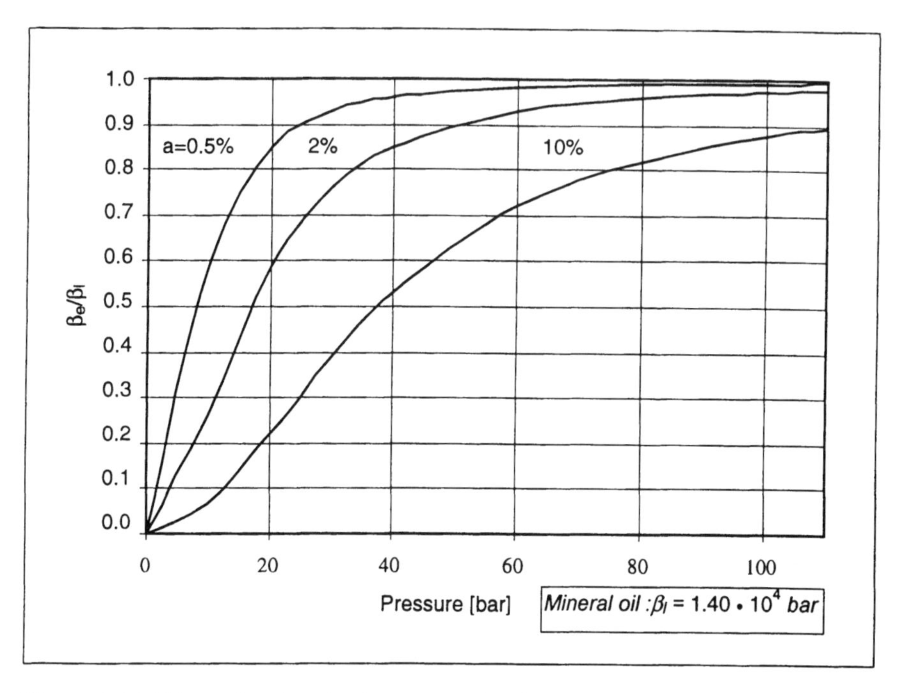

Figure 3.14 Effect of undissolved air/mineral oil ratio on bulk modulus

The relative high specific heat of water is related to the hydrogen bonding between adjacent water molecules and means that water per-unit-mass can store relatively much more energy (heat) than most other substances.

In Figure 3.15 the specific heat of water is shown, as a function of temperature and pressure.

C_P varies only little in the temperature range 0–100°C and the pressure range 1-300 bar. For lower pressures C_P has a slight minimum at 40–50°C and decreases somewhat for increasing pressures [12].

The thermal conductivity of a substance is its property to transfer energy (heat) arisen from temperature differences between adjacent parts of the substance, for instance due to power losses in the system.

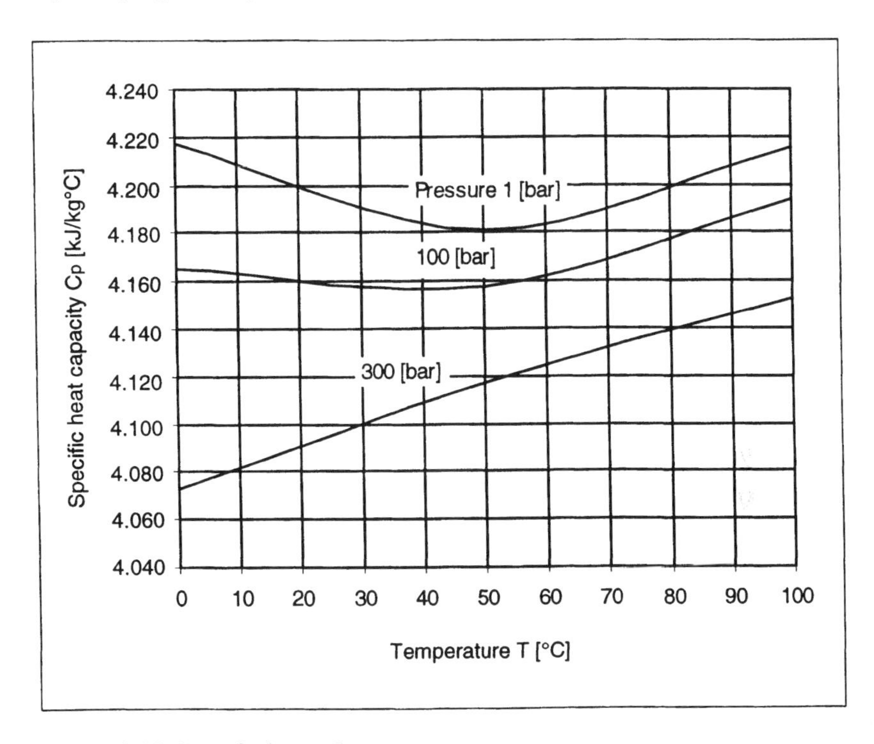

Figure 3.15 Specific heat of water

The thermal conductivity, λ of water increases with increasing temperature and pressure [12]. A graph illustrating this relationship for the temperature range 0–100°C and the pressure range 1–500 is shown in Figure 3.16.

For comparison typical values of the specific heat capacity (at constant pressure) and the thermal conductivity of mineral oils are presented in Table 3.3 and Table 3.4, respectively. The specific heat increases somewhat with increasing temperature and decreases slightly with increasing pressure, for the temperature and pressure ranges indicated. The thermal conductivity decreases with increasing temperature. The effect of pressure is of no practical significance.

In Table 3.5 typical average values for specific heats (at constant pressure) and thermal conductivities for water and mineral oil are summarised.

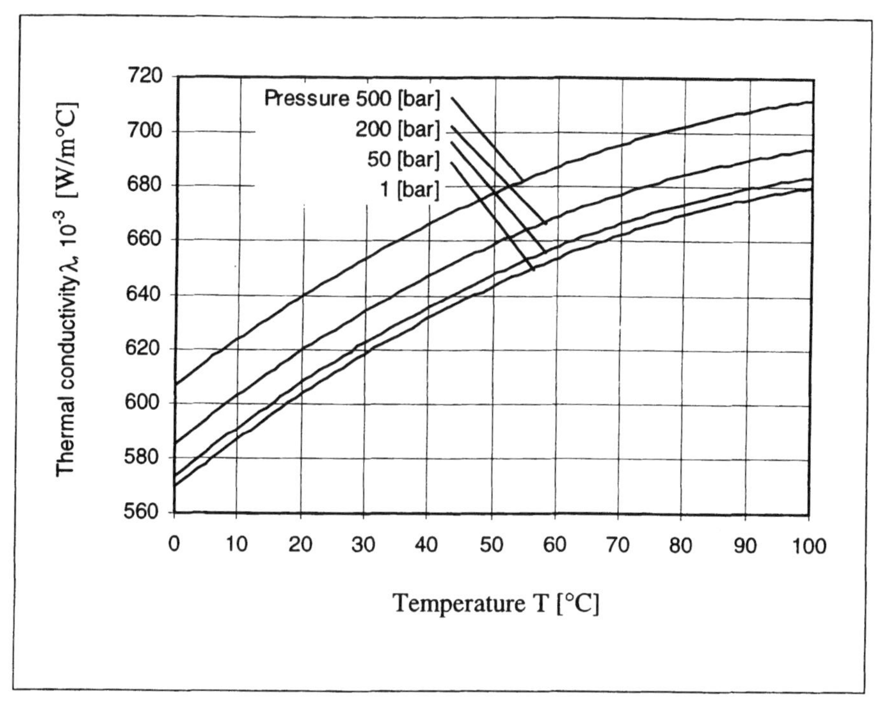

Figure 3.16 Thermal conductivity of water

As can been seen, the same mass of water can store ~2.2 times the energy (heat) that can be stored by mineral oil and water transmits energy (heat) ~5 times faster than mineral oil.

Table 3.3 Specific heat, C_P at constant pressure (Shell Tellus Oils) [5]

	C_P[kJ/kg °C]		
Temperature [°C]	Atmospheric pressure	Pressure 690 [bar]	Pressure 1380 [bar]
10	1.842	1.800	1.758
37.7	1.926	1.884	1.842
65.6	2.031	1.968	1.926
93.3	2.135	2.072	2.010
121	2.219	2.156	2.114

Table 3.4 Thermal conductivity, λ (Shell Tellus Oils) [5]

Temperature [°C]	λ[watt/m °C]
10	0.133
37.7	0.131
65.6	0.129
93.3	0.126
121	0.125

Table 3.5 Typical average values for specific heat capacity, C_P, at constant pressure and thermal conductivity, λ for water and mineral oil

	Water	Mineral oil
C_P[kJ/kg °C]	4.180	1.90
λ [W/m °C]	0.600	0.12

When designing water hydraulic systems advantages of these properties should be taken during the layout of the system and the dimensioning of the water tank (size of reservoir), cooling and heating facilities in order to optimise a proper heat balance.

f. Cavitation and aeration

In hydraulic systems it may happen that the fluid does not at some time fill out the enclosed volume of the circuit, and cavities form in the fluid. The term cavitation means the formation or the collapse of such cavities in the fluid. The cavities are fluid vapour and/or air bubbles. The collapse of these bubbles (implosions) takes place at regions where pressure suddenly rises, and within such zones the cavitation may cause damage to the surrounding component walls and/or the moving members of valves, pumps, etc. Cavitation is usually associated with an excessive noise in pumps, vibration in pipelines and erratic operation of motors and control valves.

Vapour cavitation occurs when the pressure drops to the vapour tension of the fluid at the existing temperature, when the fluid boils and fills out the cavities with vapour. Such pressure drops may arise for instance in pump suction lines, if they are too small (fluid velocity too large) or contain restrictions like strainers, sharp bends etc, or the suction level is too high.

As an example, the pressure relationship in the suction line of a hydraulic water pump is illustrated in Figure 3.17. The safety margin for cavitation should be approx 0.8 bar.

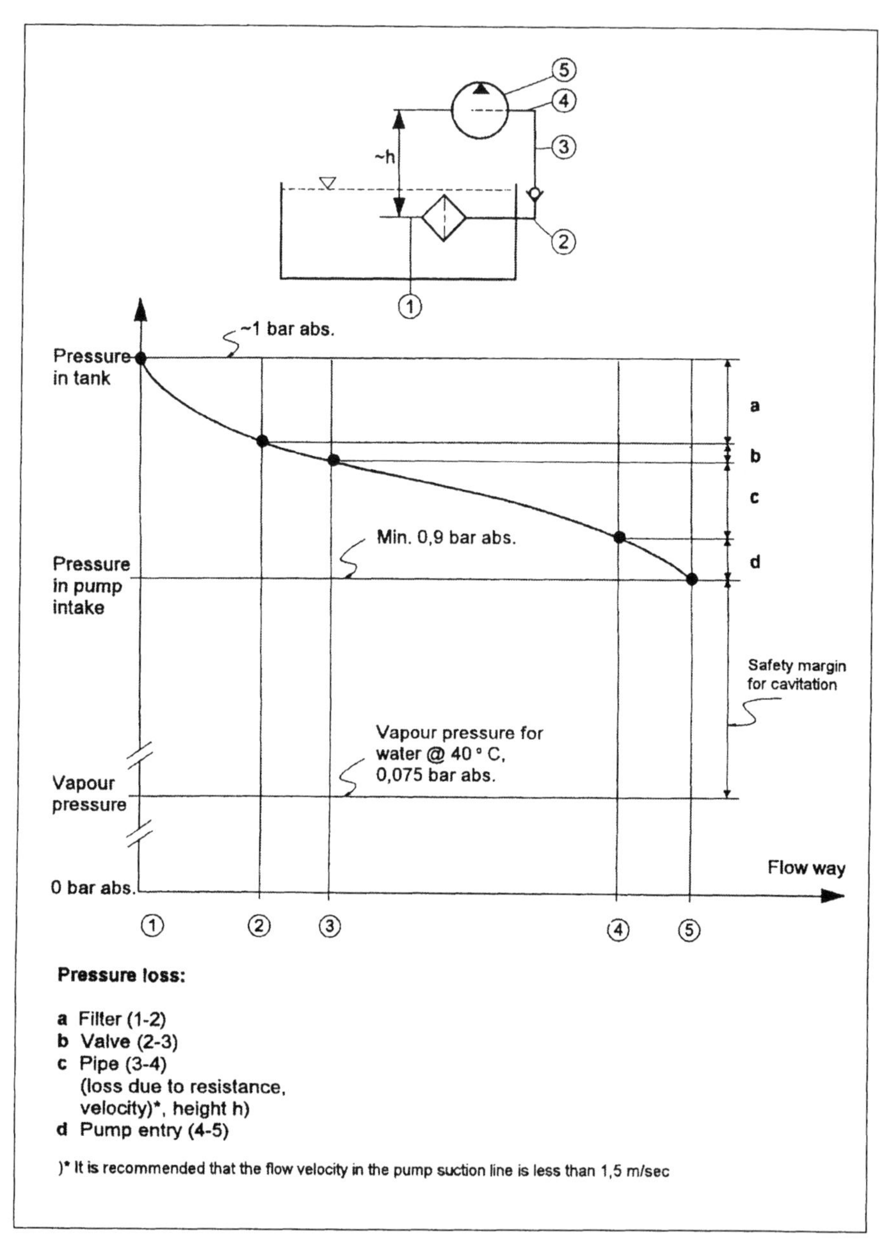

Figure 3.17 Example of pressure losses in a water pump suction line [7]

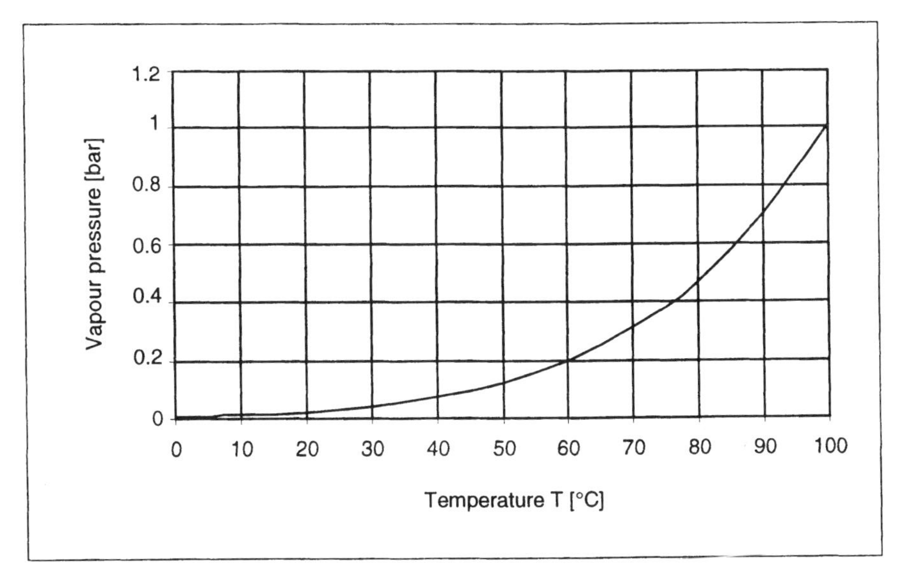

Figure 3.18 Vapour pressure of water as a function of temperature

Vapour cavitation is primarily a problem in water hydraulic systems, due to the relative high vapour pressure of the water, see Figure 3.18 Water hydraulic systems normal temperature operation interval should be limited to 3–50°C in order to avoid vapour cavitation. In some cases it may therefore be necessary to regulate the temperature of the hydraulic fluid by cooling or heating.

For comparison the vapour pressure for water and a mineral oil is illustrated in Figure 3.19. Vapour pressure increases with increasing temperature. As can be seen however, vapour pressure is not posing any cavitation problem in hydraulic systems using mineral oil due to its extreme low vapour pressure even at relatively high operational temperatures, e.g. ~$2x10^{-5}$ bar @70°C.

Air cavitation occurs when undissolved air fills out the above-mentioned cavities in the fluid. All hydraulic fluids contain dissolved air. The amount of dissolved air depends upon the type of fluid, the temperature and the pressure in the system. Dissolved air in the fluid does not pose a problem, as long as it is dissolved. However, any increase in temperature or decrease in pressure will cause air to come out of the solution. Undissolved air may also be entrained in the fluid reservoir or by leaks from fittings in the pump suction lines, and cylinder and motor return sides. Note, in order to reduce the amount of undissolved air in a system a flushing with the pressure medium and a deareation

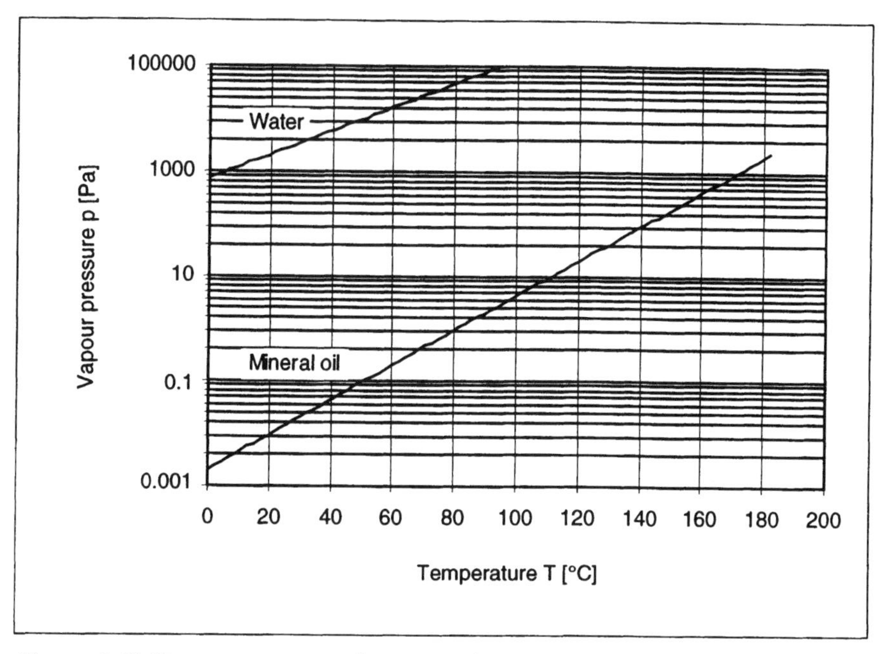

Figure 3.19 Vapour pressure of water and a mineral oil as function of temperature

should be carried out when the system has been out of operation for some time, or has been disassembled for repair. Undissolved air may cause cavitation on the suction side of hydraulic pumps (Figure 3.17) and at places in the hydraulic system where pressure drops occur, e.g. in valves. On the pressure side, air bubbles may collapse and give rise to damages like vapour cavitation (breakdowns, noise and erosion). Further, undissolved air on the pressure side may drastically reduce the effective bulk modulus, β_e (see Figure 3.13 and Figure 3.14).

In accordance with Henry's law the amount of dissolved air in a hydraulic fluid is directly proportional to the absolute pressure above the surface of the fluid and decreases with increasing temperature.

The amount of dissolved air is usually expressed by using Bunsen's absorption coefficient, b. This coefficient is defined as the volume of air measured at 0°C and at atmospheric pressure that will dissolve in a unit volume of the hydraulic fluid at a partial pressure equal to the atmospheric pressure.

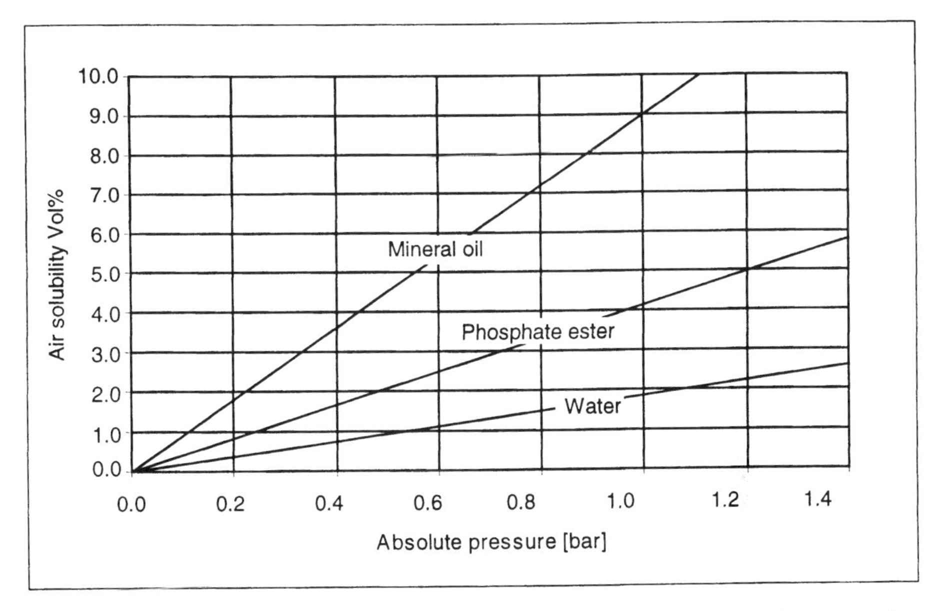

Figure 3.20 Air solubility for some hydraulic pressure media as function of absolute pressure

In Figure 3.20 the air solubility for some hydraulic fluids is shown as a function of absolute pressure. Bunsen's absorption coefficient for water is b = 0.02 (2%) and for mineral oil is b = 0.09 (9%) [7]. Note that if the pressure is doubled, the equivalent free volume of air that can be dissolved is also doubled, and so on.

3.2.2 Transport properties

a. Viscosity

Viscosity is a measure of the internal friction in a fluid, when one layer of the fluid is moved relative to another layer, thereby causing friction between the fluid molecules when the layers are moving with uneven velocities. Viscosity is one of the most important properties of a hydraulic fluid and it has a great influence on the performance of a hydraulic system and its components.

If the viscosity is high the hydraulic component efficiencies will be low because of the power loss (pressure drop) to overcome fluid friction during fluid flow. On the other hand, if the viscosity becomes low both external and internal leakage will increase and thereby cause an increasing power loss. It is therefore important, during the design of the hydraulic components, to ensure an optimal balance between lubrication requirements and the mechanical and volumetric efficiency, when the pressure medium and its viscosity interval have been selected [7].

When the temperature increases or the pressure decreases, the intermolecular cohesive forces decrease. A precise statement of the viscosity of a pressure medium therefore requires a reference to the prevailing temperature and pressure.

The following definition is attributable to Isac Newton. Two large, parallel plates at a small distance, B apart are illustrated in Figure 3.21. The space between the plates is filled with a fluid.

The lower plate is assumed to be stationary and of infinite area. The other plate, the upper one, is assumed to be of area, A and to move with a constant velocity as shown. The fluid in contact with the upper plate adheres to the plate and moves with the velocity, V. Likewise, the fluid adheres to the fixed plate and has zero velocity. Each intermediate layer will move with a velocity v directly proportional to its distance, b apart from the stationary plate. In order to move the upper plate with a constant velocity, V (see Figure 3.21) a tangential force, F must be acting upon the upper plate.

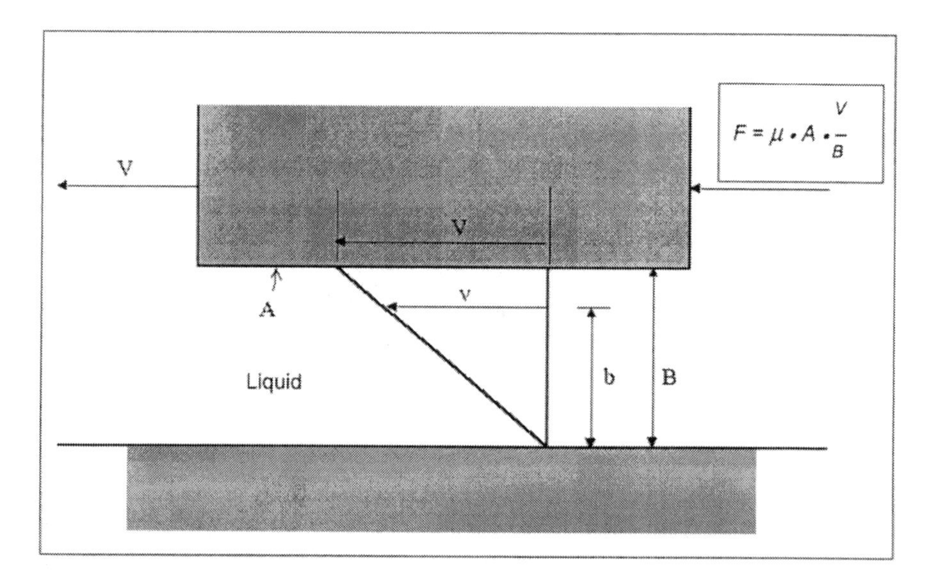

Figure 3.21 Definition of dynamic viscosity

A shear stress, τ acting on the fluid and causing the relative movement of the layers is expressed by:

$$\tau = \frac{F}{A} \tag{3.26}$$

The rate of shear, the velocity gradient, G is expressed by:

$$G = \frac{v}{b} = \frac{V}{B} \tag{3.27}$$

Newton introduced a coefficient μ, the absolute or dynamic viscosity and derived the viscosity law by:

$$F = \mu \cdot A \cdot G = \mu \cdot A \cdot \frac{V}{B}$$

where:

$$\mu = \frac{\frac{F}{A}}{\frac{V}{B}} = \frac{shear\ stress}{velocity\ gradient} \tag{3.28}$$

If the dynamic viscosity, μ, taken at any given temperature, is independent of the shear rate the fluid is said to be Newtonian. If it varies with the shear rate it is said to be non-Newtonian.

Most fluids are Newtonian, except water-in-oil emulsions and some greases.

The concept of absolute or dynamic viscosity is often used by physicists. In engineering practice the quantity ν, kinematic viscosity, is used. The kinematic viscosity is defined by the ratio:

$$\nu = \frac{\mu}{\rho} \tag{3.29}$$

where ρ is the mass density.

Since it has not been possible through direct methods to measure the dynamic and kinematic viscosity in physical units (e.g. SI-units, cgs system, English or metric gravitational systems) a series of instruments, viscometers, for indirect measurement of viscosity has been developed. Such viscometers have been the basis for defining various scales and units of viscosity.

Three standard types of short tube efflux viscometers are in use in industry for measuring empirical viscosity: In USA the Saybolt Universal and Saybolt Furol (Saybolt Universal seconds), in UK the Redwood (Redwood seconds) and in Europe the Engler (degrees °E). Tables for converting kinematic viscosity, ν, between the above empirical units can be found in engineering literature. It is important that such conversion is carried out at the same temperature.

The usual units for dynamic and kinematic viscosity are summarized in Table 3.6.

Table 3.6 Usual physical units for viscosity

Dynamic viscosity, μ
$1\,\text{Ns/m}^2 = 1\,\text{kg/ms} = 1\,\text{Pa s}$
$1\,\text{Poise} = 1\,\text{g/cms} = 1.0000 \cdot 10^{-1}\,\text{Pa s} = 100\,\text{Centipoise (cP)}$
$1\,\text{lb/fts} = 1.4882\,\text{Pa s}$
$1\,\text{Reyn (R)} = 10^6\,\mu\text{R (microreyn)} = 1\,\text{lbfs/in}^2 = 6.895 \cdot 10^6\,\text{cP}$
Kinematic viscosity, ν
$1\,\text{m}^2/\text{s} = 10^4\,\text{cm}^2/\text{s} = 10^4\,\text{Stoke (St)} = 10^6\,\text{Centistoke (c St)}$
$1\,\text{c St} = 0.01\,\text{St} = 1\,\text{mm}^2/\text{s}$
$1\,\text{ft}^2/\text{s} = 9.2903 \cdot 10^{-2}\,\text{m}^2/\text{s}$

In Figure 3.22 and Figure 3.23 the dynamic viscosity, μ and the kinematic viscosity, ν of water are shown as a function of temperature in the range 0–100°C and pressure in the range 1–200/300 bar, respectively. The viscosity for water in comparison with other liquids does not show a dramatic variation with temperature, and very little change as a function of pressure [2], [12].

Considering the normal temperature interval for operation of a water hydraulic system, the kinematic viscosity will decrease from 1.685 cSt @ 3°C and 1 bar (abs) to 0.5537 cSt @ 50°C and 1 bar (abs), i.e. a drop in viscosity of approx 3 times or (in average) a decrease of 0.024 cSt per °C.

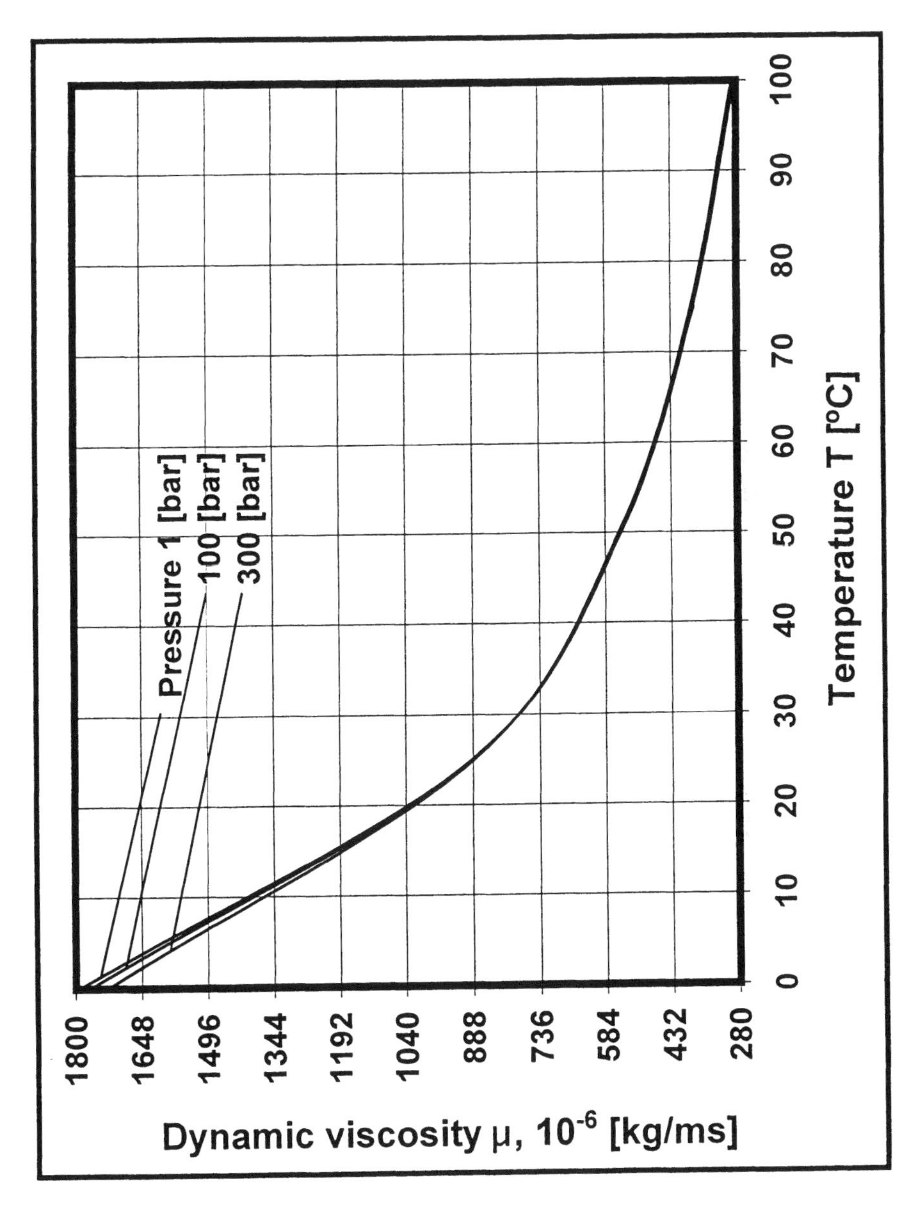

Figure 3.22 Dynamic viscosity of water as a function and temperature and pressure

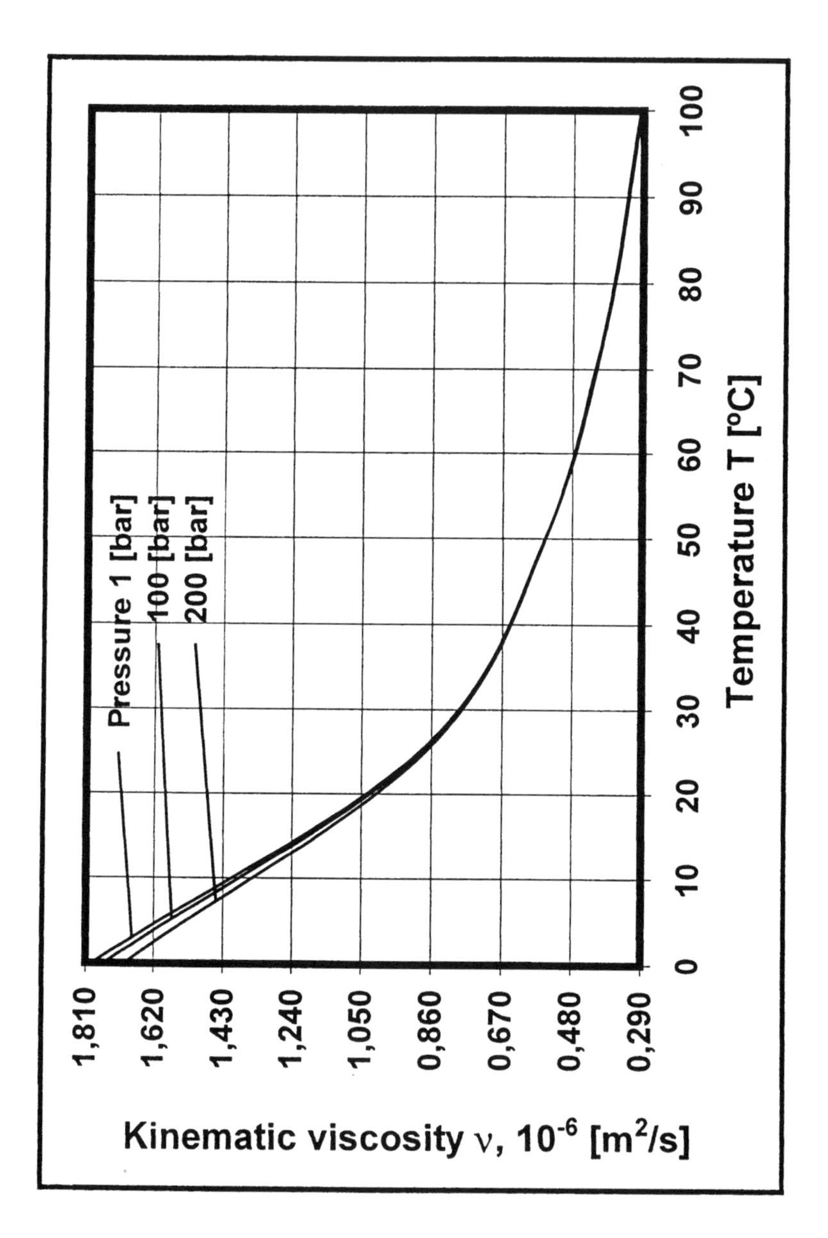

Figure 3.23 Kinematic viscosity of water as a function of temperature and pressure

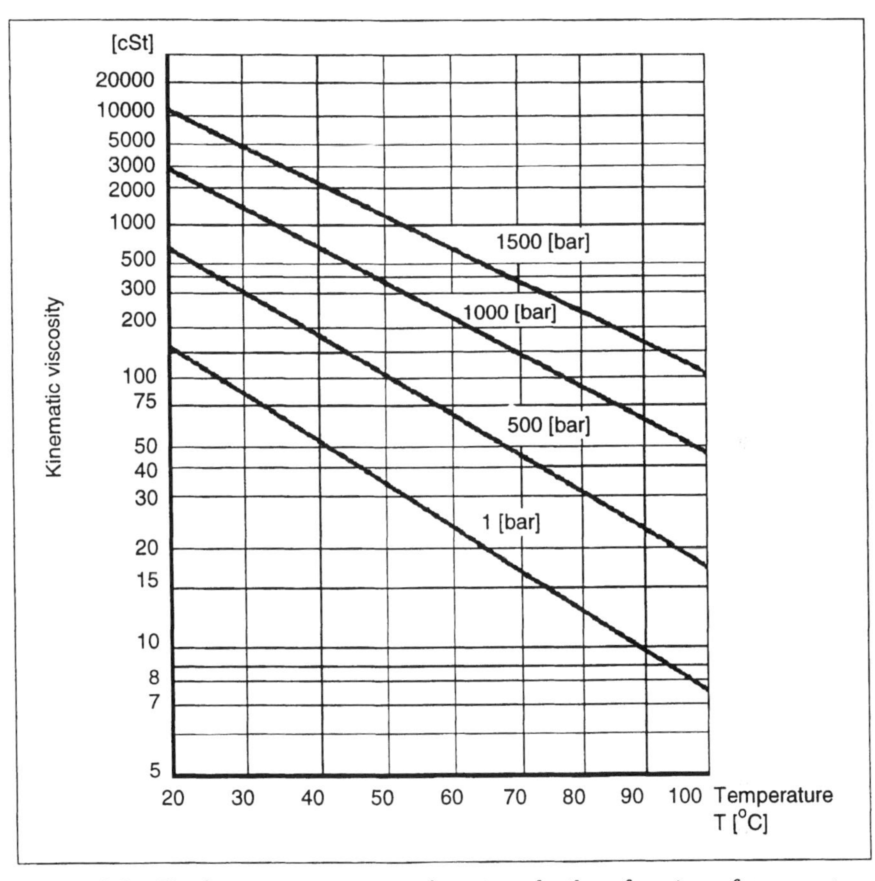

Figure 3.24 The kinematic viscosity of a mineral oil as function of temperature and pressure

For comparison, in Figure 3.24 the kinematic viscosity, ν for a mineral oil is shown as a function of temperature in the range 20–100°C and of pressure in the range 1–1500 bar [8]. Considering a temperature in the range 34–90°C for operation of an oil hydraulic system using said mineral oil as pressure medium, the kinematic viscosity will decrease from 75 cSt @ 34°C and 1 bar (abs) to 9.6 cSt @ 90°C and 1 bar (abs), i.e. a drop in viscosity of approx 8 times or (on average) a decrease of 1.17 cSt per °C.

Also note the kinematic viscosity, ν of mineral oil's sensitivity to pressure: At 50°C the viscosity, ν will increase ~ 0.7 cSt per bar.

The low viscosity of water usually leads to smaller pressure drops in circuits and lower power losses compared to mineral oil under comparable conditions.

b. Speed of sound

The speed of sound, w, can be derived from:

$$w = \sqrt{\frac{\beta_s}{\rho}} \tag{3.30}$$

where β_s is the isentropic bulk modulus and ρ is the density. As seen in (3.30) speed of sound, w depends on bulk modulus (compressibility) and density, and thus on temperature and pressure. The speed of sound in pure water as a function of temperature in the range 0–100°C and of pressure in the range 1-300 bar is shown in Figure 3.25. It has a maximum at approx 70°C at 1 bar (abs). The maximum shifts somewhat towards higher temperature for increasing pressure [14].

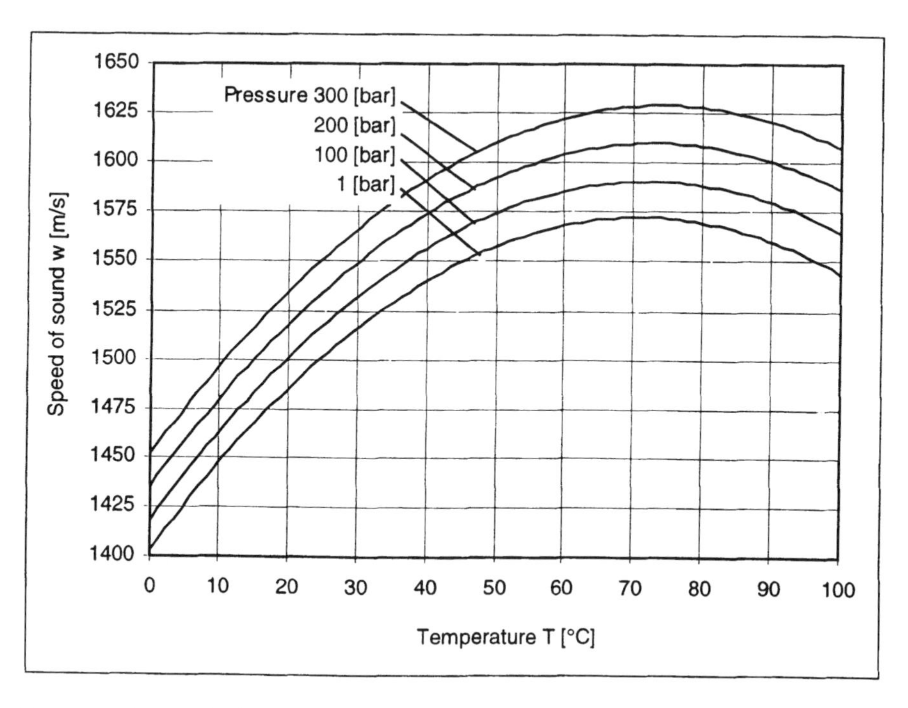

Figure 3.25 Speed of sound in water

Water hammer is the term used to express the resulting pressure shock caused by the sudden decrease of the fluid flow in a hydraulic system, for instance when a valve suddenly closes or a piston hits the bottom of a cylinder. The velocity (celerity) of the resulting pressure wave can be derived by equation (3.30) when using the effective bulk modulus, β_e instead of the isentropic bulk modulus, β_s. For comparison the speed of sound in a mineral oil is shown in Figure 3.26 in the pressure range 0-1400 bar and the temperature range 10 – 121 °C.

In Figure 3.25 and Figure 3.26 it is seen that the speed of sound, w increases for both water and mineral oil for increasing pressure [5]. For water, the speed of sound increases with increasing temperature up to a maximum at ~ 70-75°C and it then falls down with further increases in temperature. For mineral oil the speed of sound decreases with increasing temperature.

Corresponding to normal operational ranges in temperature, T and pressure for the use of mineral oil and water as hydraulic pressure media, the minimum and maximum values for speed of sound, w are shown in Table 3.7.

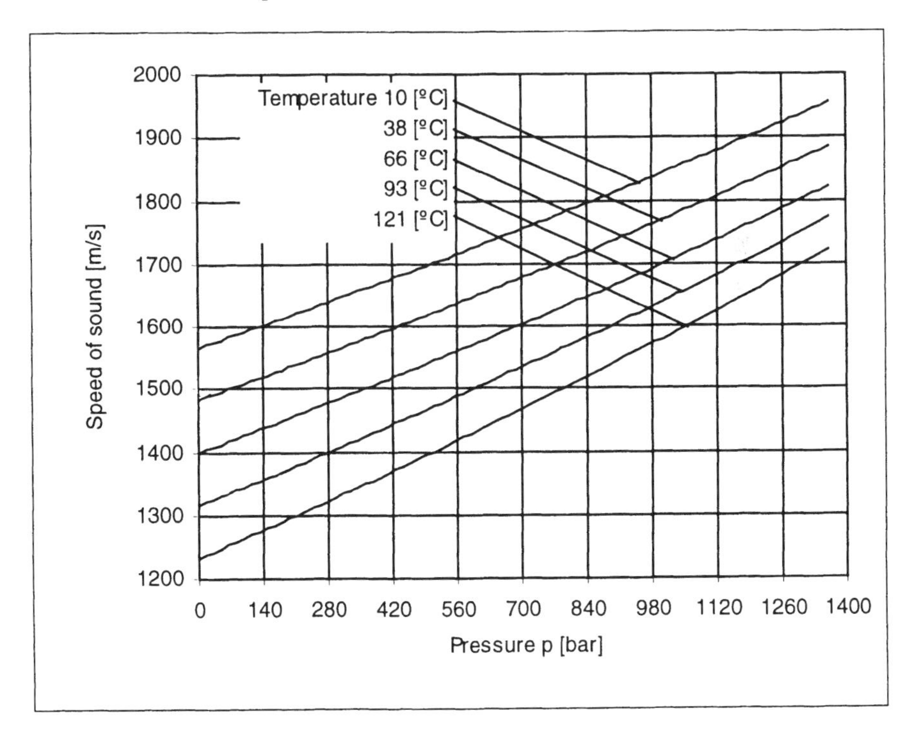

Figure 3.26 Speed of sound in a mineral oil

Table 3.7 Minimum and maximum operational values for speed of sound, w in pure water and a mineral oil

Pressure	1 [bar (abs)]		300 [bar (abs)]	
	T [°C]	W [m/s]	T [°C]	W [m/s]
Water	3	1405	3	1458
	50	1558	50	1610
Mineral oil	34	1483	34	1561
	90	1317	90	1410

Note that water hammer effects for oil hydraulic systems generally are less severe due to the lower bulk modulus, lower density and higher viscosity of mineral oils compared with water.

3.2.3 Electrical properties

a. Dielectric constant

The dielectric constant of a substance indicates how many times the capacity of a plate condensator increases when the space between the condensator plates is filled out with the substance. In Figure 3.27 a graph displaying the dielectric constant, D for water is shown as a function of temperature in the range 0-100°C and pressure in the range 1-2000 bar. The dielectric constant decreases ~ 1/3 of its value of 87 at 0°C and 1 bar (abs) to the value of 55.4 at 100°C and 1 bar (abs) [9]. For comparison some values for the dielectric constant, D for various materials are shown in Table 3.8.

Table 3.8 Some values for the dielectric constant at atmospheric pressure

Material	D
Atmospheric air	1.000
Teflon	2
Benzene	2.2
Paraffin oil	4-5
Water 25°C	78.5
Hydrogen cyanide (HCN)	116

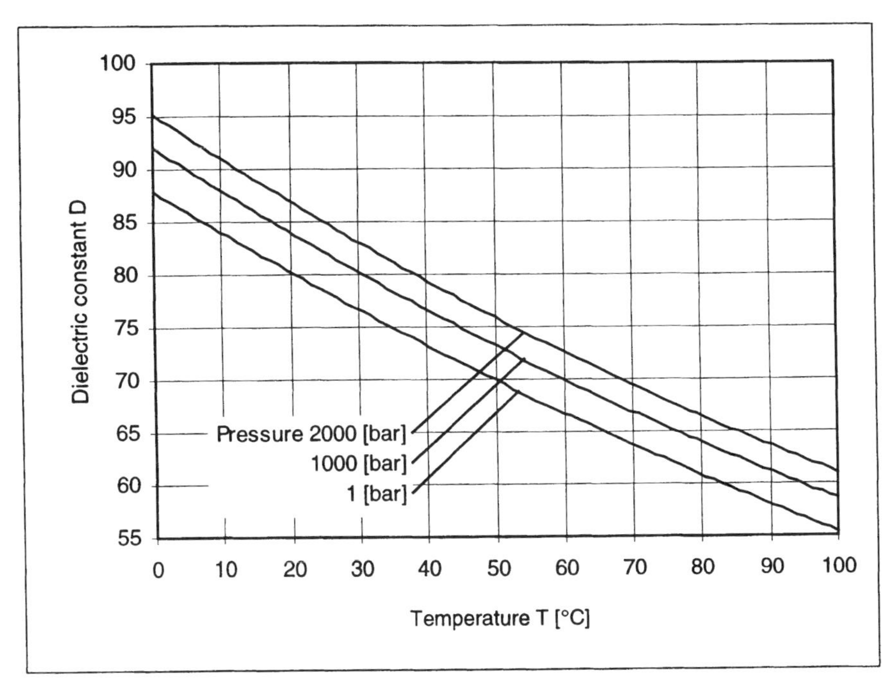

Figure 3.27 Dielectric constant for water

The relative high dielectric constant, D for water is caused by a co-operative alignment of molecules in the hydrogen-bonded network. The dielectric constant, D is a key quantity for interpreting solute-solvent interactions and explains why water is capable of dissolving a wide variety of salts, acids and bases than most other liquids. Its change with temperature and pressure makes drastic changes in the solvent properties of water.

b. Electrical conductance

Values of the specific conductance χ of $(3 - 1) \cdot 10^{-8}$ $[\Omega^{-1}]$ at 18°C have been measured in accurate measurements for pure water.

3.2.4 Other properties

a. Surface tension

The term surface tension is used loosely to identify the apparent stress in the surface of the liquid. The free surface layer of a liquid behaves like a stretched membrane.

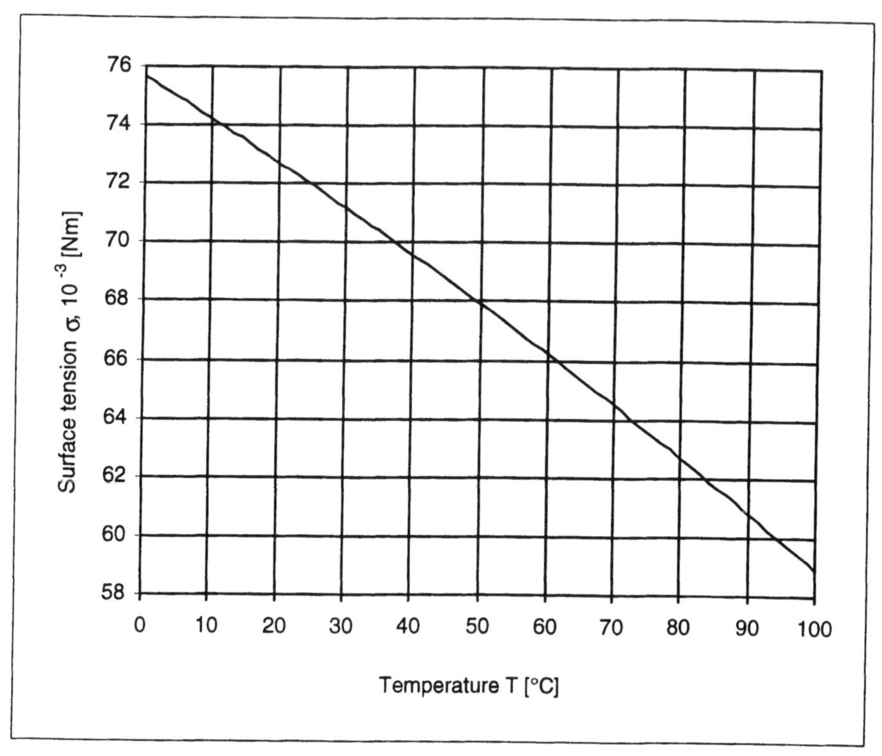

Figure 3.28 Surface tension in water

The surface tension originates from the cohesive forces on the molecules of the liquid. A molecule in the interior of the liquid is under attractive forces in all directions and the vector sum of these forces is zero. A molecule at the surface of the liquid is under a downward cohesive resulting force, which is perpendicular to the surface and caused by the asymmetrical effect from the molecules in the surrounding hemisphere.

When the surface of the liquid meets a fixed wall, its form will generally be different from the form farther away from the wall. This is caused by the adhesive forces on the molecules of the liquid from the material of the wall. The form of the surface becomes perpendicular to the combined cohesive and adhesive forces.

In Figure 3.28 the surface tension of water is shown as a function of temperature in the range 0–100°C. The surface tension decreases when temperature increases. The unit of surface tension is normally [J/m^2] which is equivalent to force per unit length, i.e. [N/m]. Surface tension can be regarded as a "cutting force" in the liquid surface.

3.3 Examples

3.3.1 Example

The volume of pure water changes with temperature. The volume change, ΔV_T can be calculated by the following equation:

$$\Delta V_T = V \cdot \alpha_P \cdot \Delta T \tag{3.31}$$

where:

V is the initial volume
α_P is the isopiestic coefficient of thermal expansion
ΔT is a temperature increase

The volume of pure water changes with pressure. The volume change, ΔV_P can be calculated from equation (3.11) slightly rewritten:

$$\Delta V_P = V \cdot \frac{1}{\beta_T} \cdot \Delta P \tag{3.32}$$

where:

V is the initial volume
β_T is isothermal tangent bulk modulus
ΔP is a pressure increase

Assuming ΔV_T and ΔV_P are numerically equal and using equations (3.31) and (3.32) the following relation can be derived:

$$\frac{\Delta P}{\Delta T} = \alpha_P \cdot \beta_T \tag{3.33}$$

For the purpose of illustration, the following numerical values are substituted into the equations (3.31) and (3.32):

V = $0.2\ m^3$
α_P = $0.35 \cdot 10^{-3}\ [°C]^{-1}$ (average value in the temperature range 20–70°C, see Figure 3.7)
ΔT = 50°C
β_T = $23.5 \cdot 10^3$ bar (average value in the temperature range 20–70°C and the pressure range 100-300 bar, see Figure 3.10).

Assuming ΔV_T and ΔV_p are equal the results become:

$$\Delta V_T = 0.2 \cdot 0.35 \cdot 10^{-3} \cdot 50 = 3.5 \cdot 10^{-3}\, m^3 = 3.5\, \ell.$$

$$\Delta P = \frac{\Delta V_P \cdot \beta_T}{V} = \frac{3.5 \cdot 10^{-3} \cdot 23.5 \cdot 10^3}{0.2} = 411.25\ bar$$

i.e. a temperature increase of 50°C will increase the pressure ~ 411 bar when there is no volumetric increase!

The equation (3.33) yields the result:

$$\frac{\Delta P}{\Delta T} = 0.35 \cdot 10^{-3} \cdot 23.5 \cdot 10^3 = 8.23 \frac{bar}{°C}$$

i.e. a temperature change of 1°C results in a pressure change of 8.23 bar!

3.3.2 Example

A piston with the diameter, d = 30 mm and the length, L = 100 mm is moving concentric in a fixed cylinder with a radial clearance, B = 10 μm and a constant speed of w = 5 m/s along the centreline.

How big is the braking force, F on the piston and the flowrate, Q through the clearance? Note, that there is no pressure build up at the end of the clearance.

Using equation (3.28) and Figure 3.21 the force F can be expressed by:

$$F = \mu \cdot 3.14 \cdot D \cdot L \cdot \frac{V}{B} \tag{3.34}$$

where μ is the dynamic viscosity.

The flow rate Q can be expressed by:

$$Q=\frac{B}{2}\cdot 3.14\cdot D\cdot V \tag{3.35}$$

The following numerical values are substituted into the equations (3.34) and (3.35):

$\mu = 1040\cdot 10^{-6}\,\frac{\text{kg}}{\text{ms}}$

$D = 0.03\ \text{m}$

$L = 0.1\ \text{m}$

$V = 5\,\frac{\text{m}}{\text{s}}$

$B = 10\,\mu\text{m} = 10^{-5}\,\text{m}$

The following results are obtained:

$$F = 1.04\cdot 10^{-3}\cdot 3.14\cdot 0.03\cdot 0.1\cdot 5\cdot 10^{5} = 4.90\,N$$

$$Q=\frac{10^{-5}}{2}\cdot 3.14\cdot 0.03\cdot 5.0 = 10^{-5}\cdot 0.236\,\frac{m^3}{s} = 2.36\,\frac{cm^3}{s}$$

3.3.3 Example

In a water hydraulic system is used pure water as pressure medium. The mass density of the water ρ is at 15°C and atmospheric pressure equal to:

999.10 kg/m³, see Figure 3.4.

During the operation of the system the water temperature increases to 50°C and the pressure to 50 bar. What will be the mass density of the water?

Using equation (3.4) the mass density of water can be computed by:

$$\rho(50\,°C) = \rho(15\,°C, atm.p.)\cdot\left[1-\alpha_P(T)\cdot(50\,°C - 15\,°C)\right]$$

From Figure 3.7 it is seen that the isopiestic coefficient of thermal expansion, α_P at 15°C and atmospheric pressure equals 0.15 and α_P at 50°C and atmospheric pressure equals 0.46. Furthermore the function, α_P is approx linear from 15°C to

50°C. Therefore an average value of $\alpha_P = 0.305$ is assumed. Using this value for α_P the following can be derived:

$$\rho(50\,°C) = 999.10\left[1 - 0.000305(50\,°C - 15\,°C)\right] = 988.4\frac{kg}{m^3}$$

Rewriting equation (3.16) to:

$$\Delta V = -V_o \frac{1}{\beta_{T,\ secant}} \cdot \Delta P$$

and using equation (3.1):

$$\rho = \frac{M}{V}$$

where the mass density at atmospheric pressure equals ρ_0, the following expression for ρ can be derived:

$$V = \Delta V + V_0 = V_0\left(1 - \frac{1}{\beta_{T,\ secant}} \cdot \Delta P\right)$$

$$\frac{V}{M} = \frac{1}{\rho} = \frac{1}{\rho_0}\left(1 - \frac{1}{\beta_{T,\ secant}} \cdot \Delta P\right)$$

$$\rho(T, P) = \rho_0 \frac{1}{1 - \dfrac{1}{\beta_{T,\ secant}} \cdot \Delta P} \qquad (3.36)$$

By substituting numerical values in (3.36) we find:

$$\rho(50\,°C,\ 100\ bar) = \frac{988.4}{1 - 43 \cdot 10^{-6} \cdot 10^2} = 992.67\frac{kg}{m^3}$$

where $\dfrac{1}{\beta_{T,\ secant}}$ is determined from Figure 3.10

By comparison with the graph in Figure 3.4, ρ at 50°C and 100 bar is found to be equal to 992.3 kg/m^3.

3.3.4 Example

When the pressure, P of a fluid in an infinite stiff container increases an amount δP in the time, dt the volume, V of the fluid in the container decreases by an amount δV expressed by equation (3.12) when the compression process is isentropic:

$$\beta_S = -V \cdot \frac{\delta P}{\delta V}$$

The rate of fluid volume change as $dt \to 0$ can be described by:

$$\frac{dV}{dt} = -\frac{V}{\beta_S} \cdot \frac{dP}{dt} \tag{3.37}$$

As this rate of volume change is equal to the flow rate to fill the volume in time, dt it is known as the compressibility flow rate, Q_c expressed by:

$$Q_c = -\frac{dV}{dt} = \frac{V}{\beta_S} \cdot \frac{dP}{dt} \tag{3.38}$$

I.e. the rate of volume change is expressed in terms of the rate of pressure increase.

Using equation (3.38) it can be seen that the ratio of the compressibility flow rate for water, Q_{c,H_2O}, over that for mineral oil, $Q_{c,oil}$, can be expressed by:

$$\frac{Q_{c,H_2O}}{Q_{c,oil}} = \frac{\beta_{S,oil}}{\beta_{S,H_2O}} \tag{3.39}$$

For the purpose of comparison a numerical example is assumed where the fluid pressure is 300 bar and its temperature is 50°C. From Figure 3.10, Table 3.1 and using equations (3.18) and (3.19) and from Figure 3.11, the ratio (3.39) becomes:

$$\frac{Q_{c,H_2O}}{Q_{c,oil}} = \frac{19}{23.8} \approx 0.80$$

That is, for the same conditions the compressibility flow of water is ~20% smaller than for mineral oil. This illustrates that decompression shocks in water hydraulic systems may be smaller than in oil hydraulic systems.

3.3.5 Example

The velocity of a pressure wave in a liquid can be derived by equation (3.30):

$$w = \sqrt{\frac{\beta_S}{\rho}}$$

The rise in pressure, ΔP in a pipe of length, L for a sudden valve closure at a closure time, t_c where:

$$t_c \leq \frac{2L}{w} \tag{3.40}$$

is given by [4]:

$$\Delta P = \rho \cdot u \cdot w \tag{3.41}$$

where u is the flow velocity of the water prior to valve closure and ρ is the density of the liquid.

Assuming the fluid is water with a temperature of 50°C and a pressure 300 bar the speed of sound, w (from Table 3.7) will be 1610 m/s and the density, ρ (from Figure 3.4) will be ρ = 1001 kg/m³, the pressure increase for a flow velocity of 10 m/s will derived by equation (3.41) be:

$$\Delta P = 1001 \cdot 10 \cdot 1610 \; Pa = 161.2 \; bar$$

For comparison the pressure increase for a flow with mineral oil is computed assuming a temperature of 50°C and a pressure of 300 bar for the mineral oil. The speed of sound w (from Figure 3.26) is 1520 m/s and the density ρ (from Figure 3.5) is ρ = 862 kg/m³. The flow velocity prior to a sudden valve closure is 10 m/s. The pressure increase derived from equation (3.41) is:

$$\Delta P = 862 \cdot 10 \cdot 1520 \; Pa = 131 \; bar$$

It is noted that in this numerical example the pressure rise due to the instant valve closure when using water as the hydraulic medium increases 23% compared to using mineral oil as the hydraulic medium under the same conditions.

3.4 Appendix to chapter 3

Summary of characteristics of water and other hydraulic fluids [7]

Liquid	Mineral oil HLP	HFA	HFC	HFD	Bio oil (rape seed) HTG	Water
Kinematic viscosity at 50°C [mm^2/sec]	15-70	~1	20-70	15-70	32-46	0.55
Density at 15°C and 1 bar abs [g/cm^3]	0.87-0.9	~1	~1.05	~1.05	0.93	1
Vapour pressure at 50°C [bar]	$1.0 \cdot 10^{-8}$	0.1	0.1-0.15	$<10^{-5}$	?	0.12
Bulk modulus β_S at 50°C and 100 bar [N/m^2]	$1.6 \cdot 10^9$	$2.5 \cdot 10^9$	$3.5 \cdot 10^9$	$2.3\text{-}2.8 \cdot 10^9$	$1.85 \cdot 10^9$	$2.5 \cdot 10^9$
Speed of sound at 20°C and 1 bar abs [m/sec]	1500	?	?	?	?	1480
Thermal conductivity at 20°C and 1 bar abs [W/m·°C]	0.11-0.14	0.598	~0.3	~0,13	0.15-0.18	0.602
Specific heat at 20°C and constant pressure at 1 bar abs [kJ/kg·°C]]	1.89	-	-	-	-	4.20
Max Working temperature range [°C]	-20-90	5-55	-30-65	0-150	-20-80	~3-50
Flash point [°C]	210	-	-	245	250-330	-
Ignition point [°C]	320-360	-	-	505	350-500	-
Corrosion protection	Good	Sufficient	Good	Good	Very good	Poor
Environmental impact	High	High	High	High	Small	None
Relative costs for liquid [%]	100	10-15	150-200	200-400	150-300	~0.02
Usage [%]	85	4	6	2	3	<1 (at present)

References

[1] Trostmann, Erik and Peter Mads Clausen: "Hydraulic Components Using Tap Water As Pressure Medium". The Fourth Scandinavian International Conference on Fluid Power. Sept. 26-29, 1995, Tampere, Finland.

[2] *Water. A comprehensive treatise*. Edited by Felix Franks. Plenum Press, New York and London, 1972.

[3] Davis and Day: *Water: The Mirror of Science*. Heinemann Educational Books Ltd. London, Melbourne, and Toronto, 1961.

[4] Merritt, Herbert E.: *Hydraulic Control Systems*, John Wiley & Sons, New York, London, and Sydney, 1967.

[5] *Technical Data on Shell Tellus Oil*. Published by Shell International Petroleum Company, Ltd., London 1963.

[6] Oehler, Gerhard: *Hydraulic Presses*, Edward Arnold, 1968.

[7] Trostmann, Erik: *Water Hydraulics Control Technology*. Marcel Dekker, Inc., New York, Basel and Hong Kong, 1996.

[8] Trostmann, E.: *Hydraulic Control*. Lecture notes, rep. S87.51, IFS, Technical University of Denmark (In Danish), 1987.

[9] *Ullmanns Encyklopädie der technischen Chemie*, 4. Auflage, Band 24. Verlag Chemie. Weinheim, Deerfield Beach Florida, Basel, 1983.

[10] Note CS-TM j.nr. 99-174 from Danfoss A/S to the author.

[11] Watton, John: *Fluid Power Systems*, Prentice Hall, New York and London, 1989.

[12] *UK steam tables in SI units, 1970*. Thermodynamic properties of water and steam. Publ. for United Kongdom committee on the properties of steam.

[13] Hodges, P.K.B.: *Hydraulic Fluids*. Arnold, London, 1996.

[14] Haar, Lester, John S. Gallagher, and George S. Kell: *NBS/NRC Steam Tables*. Hemisphere Publishing Corporation. Washington, New York, London, 1984.

[15] *CRC Handbook of Chemistry and Physics*. Edited by David R. Lide. CRC Press. Boca Raton London, New York and Washington, D.C., 1999-2000.

[16] Trostmann, Erik and Peter Mads Clausen: "Fluid Power Control Based on Pure Tap Water", 12th Aachen Conference on Fluid Power Technology, 12-13 March, 1996. Aachen Germany.

[17] Kirk-Othmer: *Encyclopedia of Chemical Technology*, Third Edition, Volume 24, John Wiley & Sons, New York · Chichester · Brisbane · Toronto · Singapore, 1984.

4 Water chemistry

Bo Højris Olesen
Danish Technological Institute

Chemistry is the study of elements and the compounds they form. Consequently, the field of chemistry is a field of multiple aspects. Aquatic chemistry, as one of the aspects, relates to the chemistry of water and the compounds that may be included therein.

The scope of this chapter is to introduce the reader to the part of aquatic chemistry that is relevant for tap water hydraulic systems. Initially, general chemical terms are briefly described followed by a discussion of terms and conditions, particularly applicable to the chemistry of tap water.

4.1 Introduction to basic chemical terms

4.1.1 Molecules and bonds

For reasons of simplification, matter can be dealt with at three levels of detail:

- the bulk level
- the molecular level
- the atomic level

As an example, table salt (bulk level) consists of crystals of sodium chloride (molecular level) formed by a sodium and a chloride atom (atomic level). The atoms are said to be the building blocks of matter. They are also the smallest fractions in which matter can exist. The atom (see Figure 4.1) consists of a positively charged core, called the nucleus, containing a number of positively charged particles, called protons. Besides protons, the nucleus may contain a number of non-charged particles, called neutrons. Except for hydrogen, there are usually about the same number of protons and neutrons in the nucleus.

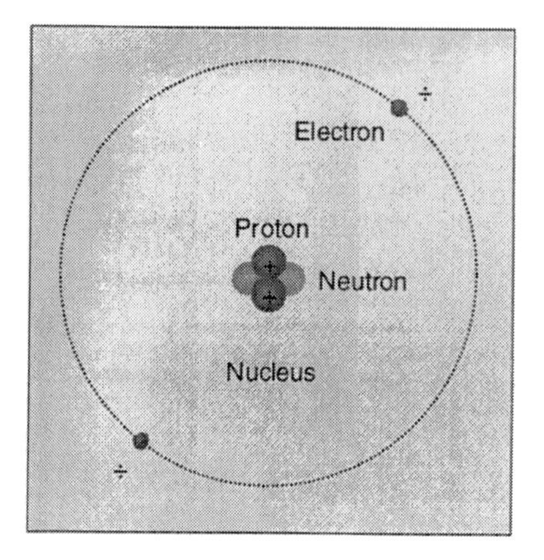

Figure 4.1 The helium atom

Surrounding the nucleus are a number of negatively charged particles, called electrons, matching the number of protons in the nucleus so that the overall charge of the atom is zero. The electrons are grouped in certain levels representing specific energies at which the negative electron can circle the nucleus without being pulled into it by the oppositely charged protons.

The phenomenon can be compared to a satellite in orbit around the earth. At a certain distance from the earth the satellite can circle the earth without falling down or without being lost into space. In reality, the electrons will not follow ideal orbits, but will move more or less randomly within a so-called electron shell. The orbit may, however, describe the distance from the nucleus where the electron is most probable to be found.

Each electron level has room for a certain number of electrons. The innermost level can contain a maximum of two electrons, the second eight, the third eighteen, the fourth thirty-two, and so on. The outermost level can, however, only contain eight electrons regardless of the number of levels below. As the atom number (the number of protons or the number of electrons in the atom) increases, the electron levels are filled from inside-out, starting with the innermost level.

The existing elements ranging from hydrogen, with an atomic number of 1, to uranium (the largest naturally occurring element) with an atomic number of 92 can be arranged in a so-called periodic table, according to the number of electron levels and the number of electrons in the outermost level.

The columns in the periodic table, the groups, represent the number of electrons in the outer level or shell. The rows, or the periods, represent the number of electron levels. A periodic table and a list of element symbols can be found on the next pages. The first group starts with hydrogen having one electron level and one electron in the outer level, followed by lithium, sodium, and potassium each with one electron in the outer electron level, but with two, three, and four electron levels respectively. The second group starts with beryllium, magnesium, and calcium each with two electrons in their outer electron level. The pattern continues to the seventh group that contains elements with seven electrons in their outer electron level and the eight group elements with eight electrons in the outer level.

The most stable situation for the atom is when the outermost electron level is filled with eight electrons or when it is empty. The elements that have their outermost electron level filled are called noble gases. The noble gases are inactive and do not react with other elements. They have already reached the highest level of stability. All the other elements have neither a filled outer electron level nor an empty one. They can, however, obtain one of these stable situations by sharing electrons with one or more other atoms. The simplest example of such an electron share is two hydrogen atoms sharing each other's single electrons.

$$H\cdot + H\cdot \rightarrow H:H \qquad (4.1)$$

Both hydrogen atoms only have one electron. By sharing electrons, both atoms experience a filled electron level. The bond between two atoms sharing electrons is called a covalent bond. The constellation of atoms sharing electrons is called molecules. Molecules can be as simple as two hydrogen atoms forming hydrogen gas ($H_{2(g)}$) or they can be complex structures like proteins containing hundreds or thousands of atoms.

Another example of a bond is the ionic bond found in e.g. table salt (sodium chloride or NaCl), where a sodium atom with one electron in the outer level bonds to a chloride atom with seven electrons in the outer level.

$$Na\cdot + \cdot\overset{\cdot\cdot}{\underset{\cdot\cdot}{Cl}}: \rightarrow Na:\overset{\cdot\cdot}{\underset{\cdot\cdot}{Cl}}: \qquad (4.2)$$

The sodium atom obtains an empty outer level as the chloride atom obtains a filled one. In the periodic table, sodium is positioned in the first group and chloride in the seventh group.

Name	Symbol	Atomic number	Name	Symbol	Atomic number
Actinium	Ac	89	Mercury	Hg	80
Aluminium	Al	13	Molybdenum	Mo	42
Americium	Am	95	Neodymium	Nd	60
Antimony	Sb	51	Neon	Ne	10
Argon	Ar	18	Neptunium	Np	93
Arsenic	As	33	Nickel	Ni	28
Astatine	At	85	Nielsborhium	Ns	107
Barium	Ba	56	Niobium	Nb	41
Berkelium	Bk	97	Nitrogen	N	7
Beryllium	Be	4	Nobelium	No	102
Bismuth	Bi	83	Osmium	Os	76
Boron	B	5	Oxygen	O	8
Bromine	Br	35	Palladium	Pd	46
Cadmium	Cd	48	Phosphorus	P	15
Calcium	Ca	20	Platinum	Pt	78
Californium	Cf	98	Plutonium	Pu	94
Carbon	C	6	Polonium	Po	84
Cerium	Ce	58	Potassium	K	19
Caesium	Cs	55	Praseodymium	Pr	59
Chlorine	Cl	17	Promethium	Pm	61
Chromium	Cr	24	Protactinium	Pa	91
Cobalt	Co	27	Radium	Ra	88
Copper	Cu	29	Radon	Rn	86
Curium	Cm	96	Rhenium	Re	75
Dysprosium	Dy	66	Rhodium	Rh	45
Einsteinium	Es	99	Rubidium	Rb	37
Erbium	Er	68	Ruthenium	Ru	44
Europium	Eu	63	Rutherfordium	Rf	104
Fermium	Fm	100	Samarium	Sm	62
Fluorine	F	9	Scandium	Sc	21
Francium	Fr	87	Seaborgium	Sg	106
Gadolinium	Gd	64	Selenium	Se	34
Gallium	Ga	31	Silicon	Si	14
Germanium	Ge	32	Silver	Ag	47
Gold	Au	79	Sodium	Na	11
Hafnium	Hf	72	Strontium	Sr	38
Hahnium	Ha	105	Sulphur	S	16
Hassium	Hs	108	Tantalum	Ta	73
Helium	He	2	Technetium	Tc	43
Holmium	Ho	67	Tellurium	Te	52
Hydrogen	H	1	Terbium	Tb	65
Indium	In	49	Thallium	Tl	81
Iodine	I	53	Thorium	Th	90
Iridium	Ir	77	Thulium	Tm	69
Iron	Fe	26	Tin	Sn	50
Krypton	Kr	36	Titanium	Ti	22
Lanthanum	La	57	Tungsten	W	74
Lawrencium	Lr	103	Uranium	U	92
Lead	Pb	82	Vanadium	V	23
Lithium	Li	3	Xenon	Xe	54
Lutetium	Lu	71	Ytterbium	Yb	70
Magnesium	Mg	12	Yttrium	Y	39
Manganese	Mn	25	Zinc	Zn	30
Meitnerium	Mt	109	Zirconium	Zr	40
Mendelevium	Md	101			

26 — Atomic number
Fe — Symbol
55.85 — Molar mass

Group

Period	IA	IIA	IIIB	IVB	VB	VIB	VIIB	---- VIII	----	----	IB	IIB	IIIA	IVA	VA	VIA	VIIA	VIIIA
1	1 **H** 1.008																	2 **He** 4.003
2	3 **Li** 6.941	4 **Be** 9.012											5 **B** 10.81	6 **C** 12.01	7 **N** 14.01	8 **O** 16.00	9 **F** 19.00	10 **Ne** 20.18
3	11 **Na** 22.99	12 **Mg** 24.31											13 **Al** 26.98	14 **Si** 28.09	15 **P** 30.97	16 **S** 32.07	17 **Cl** 35.45	18 **Ar** 39.95
4	19 **K** 39.10	20 **Ca** 40.08	21 **Sc** 44.96	22 **Ti** 47.88	23 **V** 50.94	24 **Cr** 52.00	25 **Mn** 54.94	26 **Fe** 55.85	27 **Co** 58.47	28 **Ni** 58.69	29 **Cu** 63.55	30 **Zn** 65.39	31 **Ga** 69.72	32 **Ge** 72.59	33 **As** 74.92	34 **Se** 78.96	35 **Br** 79.90	36 **Kr** 83.80
5	37 **Rb** 85.47	38 **Sr** 87.62	39 **Y** 88.91	40 **Zr** 91.22	41 **Nb** 92.91	42 **Mo** 95.94	43 **Tc** (98)	44 **Ru** 101.1	45 **Rh** 102.9	46 **Pd** 106.4	47 **Ag** 107.9	48 **Cd** 112.4	49 **In** 114.8	50 **Sn** 118.7	51 **Sb** 121.8	52 **Te** 127.6	53 **I** 126.9	54 **Xe** 131.3
6	55 **Cs** 132.9	56 **Ba** 137.3	57 **La*** 138.9	72 **Hf** 178.5	73 **Ta** 180.9	74 **W** 183.9	75 **Re** 186.2	76 **Os** 190.2	77 **Ir** 190.2	78 **Pt** 195.1	79 **Au** 197.0	80 **Hg** 200.5	81 **Tl** 204.4	82 **Pb** 207.2	83 **Bi** 209.0	84 **Po** (210)	85 **At** (210)	86 **Rn** (222)
7	87 **Fr** (223)	88 **Ra** (226)	89 **Ac**** (227)	104 **Rf** (257)	105 **Db** (260)	106 **Sg** (263)	107 **Bh** (262)	108 **Hs** (265)	109 **Mt** (266)									

Lanthanide Series*	58 **Ce** 140.1	59 **Pr** 140.9	60 **Nd** 144.2	61 **Pm** (147)	62 **Sm** 150.4	63 **Eu** 152.0	64 **Gd** 157.3	65 **Tb** 158.9	66 **Dy** 162.5	67 **Ho** 164.9	68 **Er** 167.3	69 **Tm** 168.9	70 **Yb** 173.0	71 **Lu** 175.0
Actinide Series**	90 **Th** 232.0	91 **Pa** (231)	92 **U** (238)	93 **Np** (237)	94 **Pu** (242)	95 **Am** (243)	96 **Cm** (247)	97 **Bk** (247)	98 **Cf** (249)	99 **Es** (254)	100 **Fm** (253)	101 **Md** (256)	102 **No** (254)	103 **Lr** (257)

In general, ionic bonds can be formed between elements in these two groups. Bonds can also be formed between an element in the second group and one in the sixth and so forth. Combinations are also possible in which, for example, two elements from the first group bond to one element from the sixth (e.g. H_2O) or two elements from the third group bond to three from the sixth (e.g. Fe_2O_3). In reality some elements, particularly the metals, can exist in different valences meaning that they may behave as if they came from different groups in the periodic table. For a complete description of this subject, see e.g. [Snoeyink and Jenkins, 1980] or [Stumm and Morgan, 1981].

4.1.2 Concentration and activity

In aquatic chemistry, the concentration of a certain substance in solution may be given in either mass or molar concentration. Mass concentration is defined as weight of substance per volume or per weight of solution. Mass concentrations may e.g. be given in milligrams per litre (mg/l) or in parts per million (ppm or mg/kg).

$$Concentration\,(mg/l) = \frac{weight\ of\ substance\,(mg)}{volume\ of\ solution\,(l)} \tag{4.3}$$

$$Concentration\,(ppm) = \frac{weight\ of\ substance\,(mg)}{weight\ of\ solution\,(kg)} \tag{4.4}$$

The two terms are connected through the density of the solution ($\rho_{solution}$)

$$Concentration\,(mg/l) = Concentration\,(ppm) \cdot \rho \qquad (kg/l) \tag{4.5}$$

For dilute solutions at room temperature and atmospheric pressure the density of water is close to unity (1.000 kg/l at 4°C) and the two terms of concentrations are often taken as identical. However, under extreme circumstances there may be a significant difference between the two.

Dealing with reactions among different substances, it is convenient to use molar concentrations instead of mass concentrations. The molar concentration of a compound in solution represents the number of molecules of that compound per volume of solution. Instead of writing large numbers of atoms in a solution, it is easier to use the term "mol" defined by Avogadro's number (N_A)

$$1\ mol = N_A\ molecules \tag{4.6}$$

$$N_A = 6.0238 \times 10^{23}\ molecules\ per\ mole \tag{4.7}$$

The molar concentration is thus defined as:

$$Concentration\,(mol/l) = \frac{moles\ of\ substance\,(mol)}{volume\ of\ solution\,(l)} \qquad (4.8)$$

Molar concentrations are usually given in moles per litre (mol/l) or as molarity (M). A solution containing 1 mol/l is said to have the molarity of 1 M; a solution containing 0.001 mol/l or 1 mmol/l is said to have the molarity of 1mM, etc.

Mass and molar concentrations are connected by the molar mass (M). The molar mass is the mass of 1 mol (6.0238×10^{23} molecules) of substance.

$$M = mass\ per\ mole\,(g/mol) \qquad (4.9)$$

$$Concentration\,(mg/l) = \frac{Concentration\,(mol/l)}{1000\,(mg/g)\cdot M\,(g/mol)} \qquad (4.10)$$

The molar mass depends on the elements constituting the substance in question. The periodic table contains molar masses for all elements. Molar masses of different molecules may be found by adding the molar masses at all the elements included in that molecule. The molar mass of calcium chloride ($CaCl_2$) may thus be calculated as:

$$M_{CaCl_2} = M_{Ca} + 2M_{Cl} = 40.08 + 2\cdot 35.45 = 110.98\ g/mol \qquad (4.11)$$

Concentrations, being mass or molar, are usually represented by square brackets around the compound in question.

$$[A] = Concentration\ of\ A \qquad (4.12)$$

In dilute aqueous solutions, dissolved compounds behave according to their concentration. However, as described later in section 4.3.2 , compounds in more concentrated solutions behave differently. They are said to have an activity of less than their concentration. The relation between the concentration and the actual activity of a compound is termed the activity coefficient (γ). The activity coefficient ranges from zero to a maximum of one. Braces around the compound represent activities in question.

As an example, consider adding 1 gramme (1,000 milligrammes) of sodium chloride (NaCl or table salt) to 100 millilitre (0.1 litre) of water. The mass concentration of the NaCl will then be:

$$[NaCl] = \frac{1{,}000\ mg}{0.1\ l} = 10{,}000\ mg/l = 10\ g/l$$

Assuming that the density of the water is 1 kg/l this concentration may also be given as 10,000 ppm (mg/kg). If the density of the water for some reason was slightly different, say 0.95 g/ml, the concentration given in ppm would instead be:

$$[NaCl]_{ppm} = \frac{[NaCl]_{mg/l}}{\rho_{water}} = \frac{10{,}000\ mg/l}{0.95\ g/ml} = 10{,}526\ ppm$$

The molar weights of the sodium and chloride atoms constituting the NaCl molecule can be found in periodic tables as:

$$M_{Na} = 22.9898\ g/mol$$

$$M_{Cl} = 35.453\ g/mol$$

The molar weight of the NaCl molecule can be calculated as:

$$M_{NaCl} = M_{Na} + M_{Cl} = 22.9898 + 35.453 = 58.4428\ g/mol$$

and the molar concentration as:

$$[NaCl]_{mol/l} = \frac{[NaCl]_{mg/l}}{1{,}000\ mg/g\, M_{NaCl}} = \frac{10{,}000}{1{,}000 \cdot 58.4428} = 0.1711\ mol/l$$

$$\{A\} = Activity\ of\ A \leq [A] \tag{4.13}$$

$$\{A\} = \gamma_A [A] \tag{4.14}$$

When performing chemical calculations on concentrated solutions it is essential to use activities instead of concentrations. In dilute solutions, however, the activity coefficient is close to one and the activity is thus approximately equal to the concentration.

4.1.3 Reaction, equilibrium, and steady state

Consider the reaction changing compound A to compound B:

$$A \rightleftharpoons B \tag{4.15}$$

The reaction, as written, may run in both directions either to the right from A to B or to the left from B to A, or it may run in both directions at the same time. If the reaction runs to the right, compound A is termed the reactant and compound B is termed the product and vice versa.

If the reaction takes place in a closed system, not exchanging matter with the surroundings, the summarised concentration of A and B cannot change.

$$[A+B] = [A] + [B] = constant \tag{4.16}$$

At a certain concentration of A and B the concentrations do not change with regard to time and the reaction (see equation 4.15) is said to be at equilibrium. The ratio between the concentration of the two compounds at equilibrium is termed the equilibrium constant (K).

$$K_{AB} = \frac{[B]}{[A]} \tag{4.17}$$

Theoretically, the equilibrium constant is defined as a ratio between activities; thus (e.g. equation 4.17) is an approximate equation limited to dilute solutions. The equilibrium constant itself may depend on e.g. temperature and pressure according to the nature of the reaction.

Knowing the equilibrium constant of the reaction, the concentration of the two compounds A and B may be determined as:

$$[A] = \frac{[A+B]}{1+K_{AB}} \tag{4.18}$$

$$[B] = [A+B] - [A] \tag{4.19}$$

Now, consider the reaction (equation 4.15) taking place in an open system exchanging matter with its surroundings.

Again the reaction may run in both directions with the reaction constants k_f and k_b for forward and backward reaction, respectively.

$$A \underset{k_b}{\overset{k_f}{\rightleftharpoons}} B \tag{4.20}$$

The rate of the reaction (v) in the two directions can be calculated as:

$$v_f = k_f \cdot [A] \tag{4.21}$$

$$v_b = k_b \cdot [B] \tag{4.22}$$

and the change in concentration with respect to time as:

$$\frac{d[A]}{dt} = -k_f[A] + k_b[B] \tag{4.23}$$

$$\frac{d[B]}{dt} = k_f[A] - k_b[B] \tag{4.24}$$

The term d[A]/dt represents the change of [A] with respect to time (t).

The exchange of matter with the surroundings in the open system is in this case simplified by a constant flow (Q) through a system with the volume (V). Please refer to Figure 4.2. The system is assumed completely mixed at any point of time.

If the flow into the system contains the substances A and B in the concentrations $[A]_0$ and $[B]_0$, the flow (J) of A and B into the system per system volume are given by:

$$J_A = t_r[A]_0 \quad \left(\frac{mass}{volume\ time}\right) \tag{4.25}$$

$$J_B = t_r[B]_0 \tag{4.26}$$

where t_r (equal to Q/V) is the retention time of the system. After a while, the system will reach a steady state where the concentrations of A and B within the system do not change.

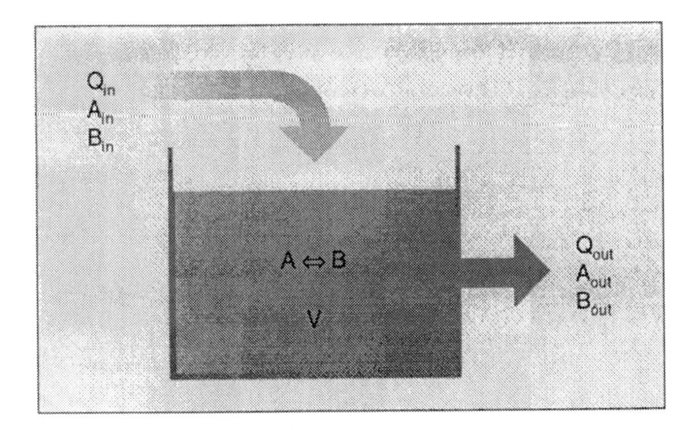

Figure 4.2 Schematic outline of balance in an open system

The steady state is not to be confused by the equilibrium that was established in the closed system, where no net reaction occurred. At steady state, the mass balance for A looks like:

$$\frac{d[A]}{dt} = J_{A,in} - J_{A,out} - consumed + produced = 0 \tag{4.27}$$

$$\frac{d[A]}{dt} = \overbrace{t_r[A]_0}^{in} - \overbrace{t_r[A]}^{out} - \overbrace{k_f[A]}^{consumed} + \overbrace{k_b[B]}^{produced} = 0 \tag{4.28}$$

and the mass balance for B looks like:

$$\frac{d[B]}{dt} = t_r[B]_0 - t_r[B] + k\ [A] - k\ [B] = 0 \tag{4.29}$$

Applying the mass balance of matter entering and leaving the system:

$$[A]_0 + [B]_0 = [A] + [B] \tag{4.30}$$

The equations (4.28 and 4.29) may be solved with respect to the steady state concentrations of A and B yielding expressions of [A] and [B] depending on the retention time, the inlet concentration and the forward and backward reaction rate:

$$[A] = \frac{t_r[A]_0 + k_b([A]_0 + [B]_0)}{k_f + k_b + t_r} \quad (4.31)$$

$$[B] = \frac{t_r[B]_0 + k_f([A]_0 + [B]_0)}{k_f + k_b + t_r} \quad (4.32)$$

The above mathematical treatment dealt with the reaction between two compounds A and B, in which one molecule of A reacted to form one molecule of B. Information about the ratio between reactants and products in a reaction is termed stoichiometry. In more general terms a chemical reaction can be written:

$$aA + bB \rightleftharpoons cC + dD \quad (4.33)$$

where A through D represents compounds and a through d are arbitrary numbers containing information on the stoichiometry of the reaction or the ratio at the different compounds in the reaction. The equilibrium constant (K) is in general terms defined as:

$$K = \frac{[C]^c[D]^d}{[A]^a[B]^b} \quad (4.34)$$

As with equation 4.17 this equation is also theoretically defined in terms of activity.

In manners similar to the above description of the reaction between A and B, it is possible to deal with reactions of higher complexity. However, it is beyond the purpose of this book to carry out such mathematical treatments.

If changes are induced to an equilibrium reaction, the equilibrium will move to whichever side reduces the stress induced by the changes in order to preserve the energy in the system. This behaviour is named "Le Chatelier's principle", after the person who firstly stated it. If, for example, the pressure of a system in which equilibrium:

$$CO_{2(gas)} \rightleftharpoons CO_{2(dissolved)} \quad (4.35)$$

is established, is increased, the equilibrium reaction will move to the right in order to reduce the volume of the system and thus the overall pressure. Another example of Le Chatelier's principle could be the introduction of heat, resulting in rearranging of equilibrium, in order to absorb the heat and minimise the temperature changes.

4.2 Chemistry of pure water

Water will result from combustion of hydrogen gas in an oxygen-filled atmosphere. The energy released by the reaction is very high, thus under normal circumstances the combustion is explosive.

$$2H_{2(g)} + O_{2(g)} \rightleftharpoons 2H_2O_{(l)} + 242\ kJ/mol\ (ca.\ 3.7\ kWh\ per\ litre) \qquad (4.36)$$

Due to the high amount of energy released during formation, water is chemically very stable (it takes at least the same amount of energy to reverse the reaction). The water molecule is composed of two hydrogen atoms bonded to an oxygen atom at an angle of 105° (see Figure 4.3). Due to the structural lack of spatial symmetry, the charge within the water molecule is not uniformly distributed and it becomes highly polar. The oxygen atom becomes slightly negative (-δ) and the two hydrogen atoms become slightly positive (+δ).

The polarity is one of the most significant properties of the water molecule. In liquid "water", the water molecules are attracted to each other due to the slightly negative oxygen and the slightly positive hydrogen. These types of bonds are called hydrogen bonds. The result in liquid water is a matrix of water molecules that are connected through hydrogen bonds (see Figure 4.3).

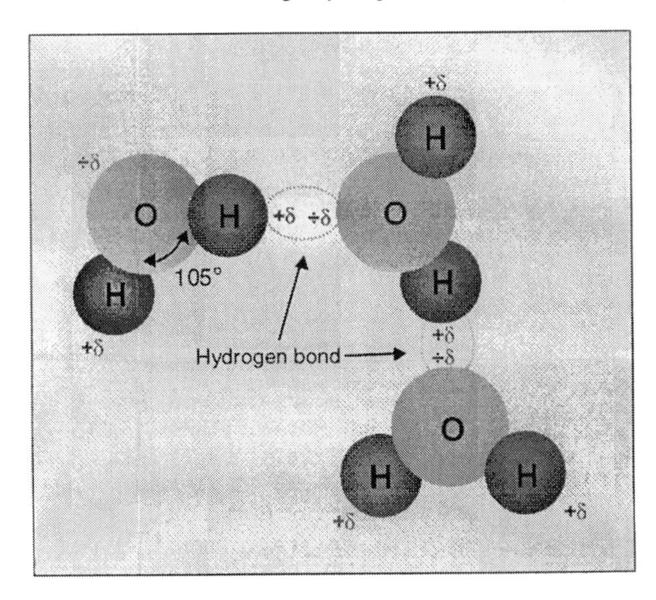

Figure 4.3 Molecular structure and spatial arrangement of water molecules

Besides the combustion reaction (equation 4.36) the only significant chemical equilibrium that exists in pure water is the autoprotolysis or the ionisation of the water itself.

$$H_2O \rightleftharpoons H^+ + OH^- \qquad (4.37)$$

The proton, when formed, will be attached to a water molecule thus, the reaction is more correctly:

$$H_2O + H_2O \rightleftharpoons H_3O^+ + OH^- \qquad (4.38)$$

This reaction is usually called the ion product of water. The degree of water ionisation is reflected in the equilibrium constant for the above reaction K_w.

$$K_w = \frac{\{H_3O^+\}\{OH^-\}}{\{H_2O\}\{H_2O\}} \qquad (4.39)$$

In pure water the activity of the water molecule is said to be unity (1 M), and for the purpose of this section it will be assumed to be so. Thus, in pure water the equation becomes:

$$K_w = \{H_3O^+\}\{OH^-\} = \{H^+\}\{OH^-\} \qquad (4.40)$$

At 25°C and atmospheric pressure, $K_w = 1.01 \times 10^{-14}\ M^2$, thus the fraction of ionised water is extremely low. The value of K_w depends on temperature and pressure. The following is an approximation to the relation between the absolute temperature (T) and K_w [*Harned and Owen, 1958*]:

$$Log(K_w) \cong -\frac{4470.99}{T(^\circ K)} + 6.0875 - 0.01706 \cdot T(^\circ K) \qquad (4.41)$$

Table 4.1 shows selected temperature values of K_w calculated using this formula.

The pressure dependence of the ionisation product constant for water has been studied under various temperature and pressure ranges. The dependency of pressure is less than the dependency of temperature. One of the most relevant studies for the conditions within tap water hydraulic systems covers the range of 5-35°C and 1-2000 bar [Whitfield, 1972]. Table 4.2 shows selected values from this study.

Table 4.1 Equilibrium constant K_w for water ionisation dependent of temperature at 1 atm pressure

Temperature [°C]	K_w [M^2]	Temperature [°C]	K_w [M^2]
0	0.12×10^{-14}	40	2.94×10^{-14}
5	0.19×10^{-14}	50	5.48×10^{-14}
10	0.29×10^{-14}	60	9.63×10^{-14}
15	0.45×10^{-14}	70	16.00×10^{-14}
20	0.68×10^{-14}	80	25.26×10^{-14}
25	1.01×10^{-14}	90	38.06×10^{-14}
30	1.47×10^{-14}	100	54.93×10^{-14}

Table 4.2 Equilibrium constant K_w for water ionisation dependent of pressure and temperature [Whitfield, 1972]

Pressure [bars]	Temperature [°C]	K_w [M^2]
1000	5	0.41×10^{-14}
	15	0.98×10^{-14}
	25	2.14×10^{-14}
	35	4.50×10^{-14}
2000	5	0.81×10^{-14}
	15	1.86×10^{-14}
	25	4.07×10^{-14}
	35	8.32×10^{-14}

4.2.1 pH

In ideally neutral water the activities of protons (H^+ or H_3O^+) and hydroxyl (OH^-) ions are equal and in equilibrium with water through the ion product of water. If the activity of protons exceeds the activity of hydroxyl ions the water is said to be acidic. In the reverse situation the water is said to be alkaline. The activity of the two species can be calculated from the equilibrium constant of water ionisation. Assuming unity water activity, atmospheric pressure, and 25°C:

$$K_w = \{H^+\}\{OH^-\} = 1.01 \times 10^{-14}\ M^2 \tag{4.42}$$

In neutral solutions:

$$\{H^+\} = \{OH^-\} = \sqrt{1.01 \times 10^{-14}} = 1.005 \times 10^{-7}\ M \tag{4.43}$$

The activity of protons is used as a measure for acidic or alkaline properties of the solution. A neutral solution at 25°C is thus a solution in which the proton activity is 1.005×10^{-7}. Acidic solutions are defined as having proton activities higher than 1.005×10^{-7} and alkaline solutions as having proton activities of less than 1.005×10^{-7}.

In order to obtain a more user-friendly scale for proton activity the term pH has been introduced and defined as:

$$pH = -log\{H^+\} \tag{4.44}$$

On the pH scale a neutral solution at 25°C and atmospheric pressure equals 6.998 or 7 and we say that the solution has a pH of 7. Due to the definition of pH, an acidic solution will have a pH below 7 and an alkaline solution will have a pH above.

Due to the fact that the equilibrium constant of water ionisation depends on temperature and pressure, so do the activity of protons in a neutral solution and thus the fix point of the pH scale. Table 4.3 and Table 4.4 show the pH in a neutral solution at different temperatures and pressures.

Table 4.3 pH in a neutral solution at different temperatures and 1 atm pressure calculated on the basis of Table 4.1

Temperature [°C]	PH
0	7.47
25	7.00
50	6.63
100	6.13

Table 4.4 pH in a neutral solution at different pressure and temperature calculated on the basis of Table 4.2

Pressure [bar]	Temperature [°C]	pH
1000	5	7.20
	35	6.67
2000	5	7.05
	35	6.54

Also the concentrations of ions can influence the ionisation product of water and thus the pH of neutrality. For seawater (about 3.5% salinity) at 25°C, pK_w has been measured to be 13.2, yielding a pH at neutrality of 6.6 [*Culberson and Pytkowicz, 1973*].

For a description of how to measure pH, see section 4.4.4 .

The logarithmic pH scale has proven quite useful in aquatic chemistry and is frequently used for concentrations of other compounds as well as equilibrium constants. Use of the scale is represented with a lower case p in front of the term of interest:

$$pK_w = -log(K_W) \quad or \quad pOH = -log\{OH^-\} \qquad (4.45)$$

4.3 Real waters

Even though pure water can be formed in nature through evaporation and condensation, it will only last a limited time before it is contaminated. Pure water as described above is thus a theoretical rather than a naturally occurring substance.

The world pool of natural waters is constituted by 97.13% seawater, 2.24% ice, 0.61% groundwater, and 0.02% freshwater [*Snoeyink and Jenkins, 1980*], containing a variety of impurities in form of dissolved and suspended matter. Table 4.5 shows the concentration of impurities in three different types of natural water.

4.3.1 Ions and solubility

Due to its polar nature water is a unique solvent of charged species. Salts dissolve well in water forming charged ions, whereas non-charged or non-polar substances like oil are practically insoluble in water. Also many gases and organic substances dissolve in water.

Solubility of inorganic solids

When a salt, say sodium chloride (NaCl), is added to water, the polar forces of the water break the molecular bonds within the salt. The salt dissociates into positively charged sodium (Na^+) and negatively charged chloride (Cl^-) ions.

$$NaCl_{(s)} \Rightarrow Na^+ + Cl^- \qquad (4.46)$$

Table 4.5 Typical constituents of different types of natural waters. [1]*[Hill, 1963],* [2]*[Snoeyink and Jenkins, 1980],* [3]*[Granat, 1972]*

Constituent [ppm]		Seawater[1]	Groundwater[2]	Rain[3]
Sodium	Na^+	10500	8.2	0.97
Magnesium	Mg^{2+}	1350	34	0.36
Calcium	Ca^{2+}	400	92	3.3
Potassium	K^+	380	1.4	0.23
Chloride	Cl^-	19000	9.6	2.0
Sulphate	SO_4^{2-}	2700	84	6.1
Bicarbonate	HCO_3^-	142	339	
Bromide	Br^-	65		
Silica	SiO_2	6.4	10	
Iron	Fe^{3+}		0.09	
Ammonium	NH_4^+			0.42
Nitrate	NO_3^-		13	2.2
Other solids		34		
Total		34571	591	16

Water molecules that are attracted by the charge of the ions form structured layers of adhered water (see Figure 4.4). Depending on the charge of the ion several layers of water can be fixed that way.

The above reaction representing the dissolution of sodium chloride is in fact an equilibrium reaction that can run in both directions. When the reaction, as written, runs to the right NaCl will dissolve into Na and Cl ions. When it runs to the left Na and Cl, ions will precipitate as solid NaCl. The position of the equilibrium or the fraction of NaCl that can be dissolved is called the solubility and is represented by the solubility product K_{so}.

$$K_{so} = \frac{\{Na^+\}\{Cl^-\}}{\{NaCl_{(s)}\}} \tag{4.47}$$

The activity of a solid is per definition taken as unity, thus:

$$K_{so} = \{Na^+\}\{Cl^-\} \tag{4.48}$$

Figure 4.5 shows solubility product constants for selected compounds.

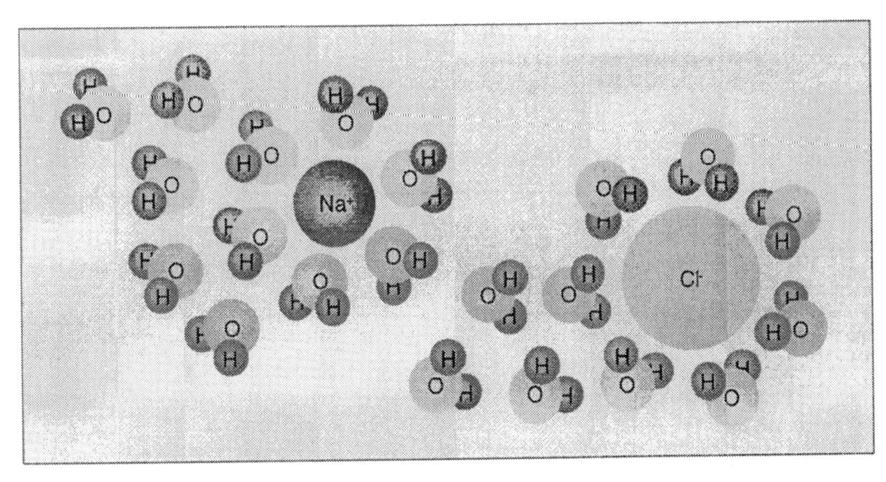

Figure 4.4 Water molecules arranged around Na+ and Cl- ions

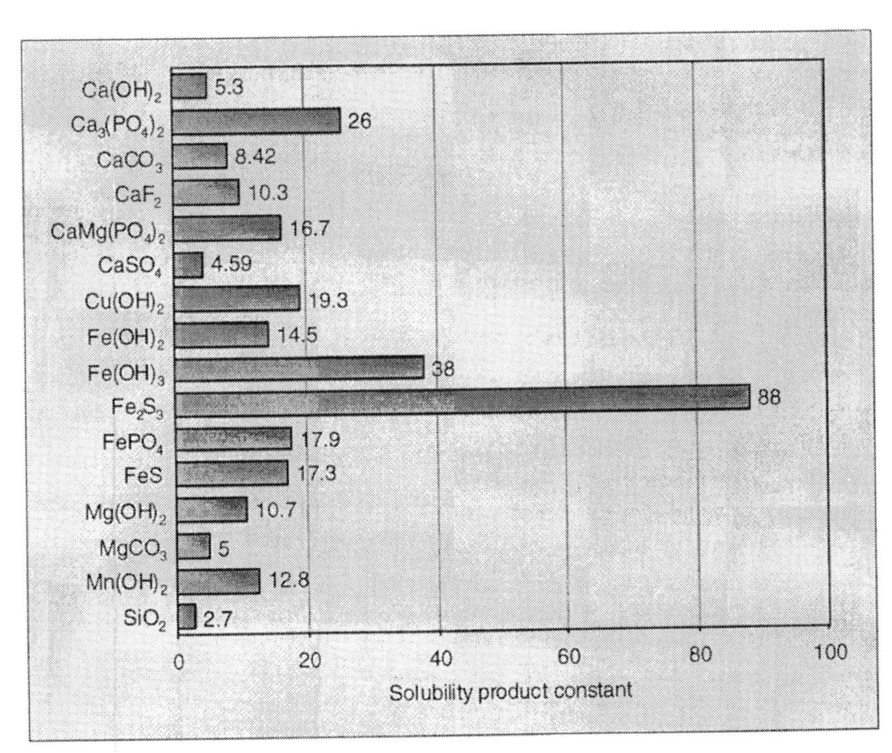

Figure 4.5 Solubility product constants for selected inorganic solid compounds at 25°C

The solubility depends, among other parameters, on temperature. Table 4.6 shows the solubility product constant for calcium carbonate at different temperatures.

Table 4.6 Solubility product constant for $CaCO_3$

Temperature [°C]	pK_{so}	Temperature [°C]	pK_{so}
5	8.35	20	8.39
10	8.36	25	8.42
15	8.37	40	8.53

Solubility of gasses in water

The solubility of gases in water is more dependent on pressure than the solubility of solid compounds.

$$CO_{2(g)} \rightleftharpoons CO_{2(aq)} \tag{4.49}$$

The reason is that the equilibrium reaction describing dissolution of gases in water contains the gaseous substance on the one side. The activity of gaseous compounds is by definition equal to the partial pressure of the gas.

$$K_{so} = \frac{\{CO_{2(aq)}\}}{P_{CO_2}} \quad or \quad \{CO_{2(aq)}\} = K_{so} \cdot P_{CO_2} \tag{4.50}$$

where P_{CO2} is the partial pressure of CO_2. The solubility of gases can also be described through Henry's law, which is basically an inverted solubility product for gases:

$$P_{CO_2} = H_{CO_2} \cdot \{CO_{2(aq)}\} \tag{4.51}$$

where H_{CO2} is the dimensionless Henry's law constant for dissolution of CO_2 in water and P_{CO_2} is the partial pressure inserted as the molar fraction of the gas.

An increase in partial pressure of a certain gas will thus increase the solubility of the gas in water. Assuming that the composition of the atmosphere within which the water is in equilibrium does not change, the partial pressure will change according to the overall pressure.

Figure 4.6 shows Henry's constant for solubility of various gases.

The temperature is also important to the solubility of gaseous compounds. Figure 4.7 shows the solubility of oxygen as an example hereof.

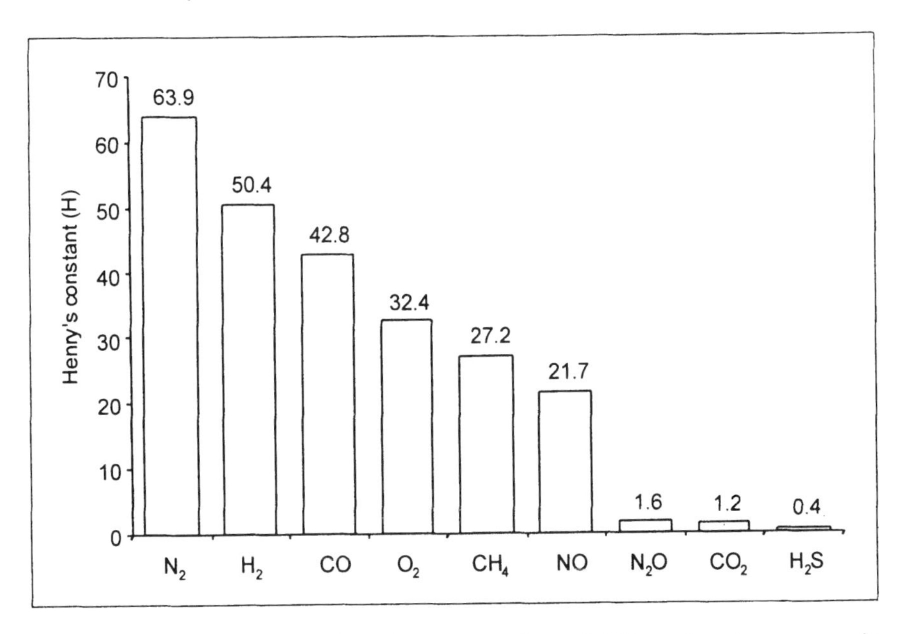

Figure 4.6 Dimensionless Henry's constant for solubility of various gases in water at 25°C

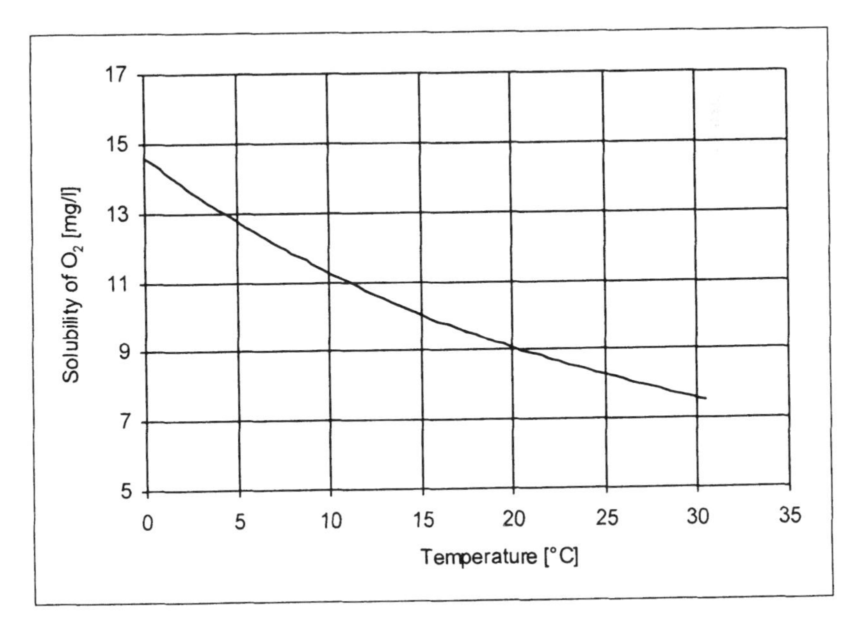

Figure 4.7 Solubility of oxygen at different temperatures at 1 atm pressure.

The pressure dependency of gas solubility can be illustrated by opening a carbonated beverage. The pressure inside the closed container is high enough to keep the carbon dioxide dissolved, but when opening the container, the pressure and thereby also the solubility of carbon dioxide drops and the gas is released as small bubbles.

This temperature dependency of gas solubility is often used to de-gas a solution. When water is heated to near boiling point, the solubility of e.g. oxygen is practically zero and all oxygen may thus be removed. As the water cools down the solubility increase again. If the water is not sealed from the atmosphere when cooling, the oxygen will thus diffuse back into the water again.

Solubility of larger molecules

When larger molecules like proteins are dissolved in water, they do not dissociate completely like salts. Proteins, for example, have certain sites within their molecular structure that can dissociate and attract water molecules. These sites are called functional groups. A carboxyl group, as an example of a common functional group, consists of:

$$R\text{-}COOH \rightleftharpoons R\text{-}COO^- + H^+ \qquad (4.52)$$

where R represents the main molecule.

Dissociation of the functional groups and adherence of water help to keep the molecule dissolved. Apart from the charged dissociated functional groups, a non-uniform charge distribution on the surface of large molecules can attract water molecules the same way water molecules are attracted to each other.

A compound with functional groups that can dissociate or with a non-uniform charge distribution on the surface, attracts the water and is called hydrophilic. A compound without functional groups and with a uniform charge distribution has no sites for interactions with water. It will repel water and is thus called hydrophobic.

Large molecules like e.g. proteins may include thousands of atoms. In such cases, it makes little sense talking about dissolution since it is only a very small part of the molecules that is dissolved (the functional groups). Instead such molecules are said to be suspended. Suspended compounds within a size range of 1 nm (0.001 μm) to 1 μm are called colloids. Above 1 μm one may talk about actual particles.

Milk is a good example of a colloid suspension. The proteins within milk are suspended, as is to some extent the milk fat. Left for a while the fat may accumulate on the surface of the milk, but the proteins stay suspended. The reason for this is that the functional groups within the proteins are dissociated and the proteins surrounded by so much adhered water that they cannot bond to each other. If acid is added to the milk or the milk is soured by micro-organisms, the pH rises and the functional groups are neutralised with protons. Since the proteins are no longer charged, there is no longer any adhered water to keep them from bonding to each other. Consequently, the milk coagulates.

Normally, colloids do not react with each other, yet they can form aggregates and flocks by bonding to each other either directly or through multiple valence ions, polymeric chains (long molecules with many charged sites or dipoles), or larger particles.

Influence of further reactions on the solubility

The solubility product only tells the ratio between the solid compound and the dissolved compound. The actual solubility, or the number of grammes that can be dissolved in water, may be different if the dissolved compound reacts further with other compounds in the water. In that way the initial concentration of the dissolved compound will decrease. Since the solubility product does not change, more material can be dissolved until equilibrium has been reached between the dissolution and the reaction of the compound.

A possible reaction could be the reaction with protons. The dissolved compound may loose or gain protons, depending on the proton activity or the pH of the water. Carbon dioxide, for example, can be dissolved in water as $H_2CO_{3(aq)}$:

$$CO_{2(g)} + H_2O \rightleftharpoons H_2CO_{3(aq)} \tag{4.53}$$

and react further with protons:

$$H_2CO_3 \rightleftharpoons H^+ + HCO_3^- \tag{4.54}$$

$$HCO_3^- \rightleftharpoons H^+ + CO_3^{2-} \tag{4.55}$$

Carbonate may also precipitate as e.g. calcium carbonate:

$$Ca^{2+} + CO_3^{2-} \rightleftharpoons CaCO_{3(s)} \tag{4.56}$$

Before the water can be said to be saturated with carbon dioxide (the highest possible amount of CO_2 has been dissolved), not only the solubility product for carbon dioxide gas has to be satisfied, but also the equilibrium constants for all of the above reactions.

4.3.2 Ionic strength and activity coefficients

The last section discussed the solubility and adherence of water around dissociated ions (see Figure 4.4). If the concentration of ions is high, the amount of adhered water is also high. Thus, there is not much "free" water available nor can the ions move freely without colliding. As an effect hereof the activity of the ions decreases as do the activity of water itself. In seawater (about 2% chloride) at 25°C it is estimated that only 2.5 mol water per kg is "free" (1 kg of pure water contains 55.5 mol water molecules) and that the activity of water is therefore decreased to 0.981 compared to unity in pure water [*Christenson and Gieskes, 1971*]. Both the concentration of ions and their charge influence activities. Actually, it is the concentration of charge that determines the activity coefficients. The concentration of charge in a water sample can be summarised in the ionic strength (μ) of the sample, which is defined as:

$$\mu = \tfrac{1}{2}\sum \left(C_i Z_i^2\right) \qquad (4.57)$$

where C_i is the concentration and Z_i the charge of species, i. The ionic strength includes both positive and negative charged ions. Depending on the charge of the dissociated ions, the same concentration of different compounds can cause different ionic strength.

Consider, for example, three one-molar (1M) solutions: sodium chloride, calcium chloride, and ferric chloride. Because of the different valences of the cations involved in the three molecules and the different corresponding number of chloride atoms, the ionic strength for the three one-molar solutions will differ as follows:

1M NaCl:	$NaCl \rightleftharpoons Na^+ + Cl^-$	$\mu = \tfrac{1}{2}\left(1\cdot 1^2 + 1\cdot 1^2\right) = 1$
1M $CaCl_2$:	$CaCl_2 \rightleftharpoons Ca^{2+} + 2Cl^-$	$\mu = \tfrac{1}{2}\left(1\cdot 2^2 + 2\cdot 1^2\right) = 3$
1M $FeCl_3$:	$FeCl_3 \rightleftharpoons Fe^{3+} + 3Cl^-$	$\mu = \tfrac{1}{2}\left(1\cdot 3^2 + 3\cdot 1^2\right) = 6$

The ionic strength correlates well with the conductivity or specific conductance since it is the charged ions in a water sample that makes it conductive. The following approximate relation between conductivity and ionic strength exists [*Russel, 1976*]:

$$\mu \approx 1.6 \times 10^{-5} \cdot specific\ \ conductance\,(\mu S/cm) \tag{4.58}$$

The ionic strength, to some extent, also correlates with the amount of total dissolved solids in a water sample [*Langelier, 1936*]:

$$\mu \approx 2.5 \times 10^{-5} \cdot total\ dissolved\ solids\ (mg/l) \tag{4.59}$$

However, since this relation does not take into account the charge of the dissolved solids, it can only be used to estimate the value of the ionic strength. The estimated values hold quite well for monovalent ions, but the errors increase significantly for higher valences.

The relation between the ionic strength and the activity coefficients of the different compounds in a solution is quite complex. However, depending on the magnitude of the ionic strength, various approximations can be used. For ionic strengths of about 0.005, which according to the above equals 312 µS/cm or about 200 mg/l total dissolved solids, or less, the DeBye-Hückel approximation may be applied with good results for activity coefficients of ions:

$$-\log(\gamma_i) = \tfrac{1}{2} Z_i^2 \sqrt{\mu} \tag{4.60}$$

where γ_i is the activity coefficient and Z_i is the charge of compound i. From this relation it will appear that the activity coefficient decreases as the ionic strength increases, but also that the activity coefficient decreases significantly as the charge of the compound grows. For higher ionic strengths, up to about 0.1 (6.25 mS/cm or about 4 g/l total dissolved solids), the Güntelberg approximation may be applied instead with a reasonable result for activity coefficients of ions:

$$-\log(\gamma_i) = \frac{\tfrac{1}{2} Z_i^2 \sqrt{\mu}}{1 + \sqrt{\mu}} \tag{4.61}$$

There exist more accurate, but yet more complex, approximations to the relation between activity coefficients and ionic strength, but the two above equations will be sufficient for calculations related to tap water hydraulics. Above an ionic strength of 0.5 there are no satisfactory equations that can describe the activity coefficient.

As an example of ionic strength and activity coefficients, consider adding 1.622 g $FeCl_3$ to 1 litre of water. Assuming the compound fully dissociates:

1.622 g $FeCl_3$ = 10 mM $FeCl_3$ = 10 mM Fe^{3+} + 30 mM Cl^-

$$\mu = \frac{1}{2}\sum\left(c_i z_i^2\right) = \frac{1}{2}\left(0.01 \cdot 3^2 + 0.03 \cdot 1^2\right) = 0.06$$

In this case the Güntelberg approximation may be applied:

$$-log(\gamma_{Cl}) = \frac{\frac{1}{2} \cdot 1^2 \sqrt{0.06}}{1+\sqrt{0.06}} \quad \Rightarrow \quad \gamma_{Cl} = 0.80$$

$$-log(\gamma_{Fe}) = \frac{\frac{1}{2} \cdot 3^2 \sqrt{0.06}}{1+\sqrt{0.06}} \quad \Rightarrow \quad \gamma_{Fe} = 0.13$$

Now, multiplying activity coefficients and concentrations, we get the actual activity of the dissociated ions:

$$\{Fe^{3+}\} = \gamma_{Fe}[Fe^{3+}] = 0.13 \cdot 0.01 = 1.3mM$$

$$\{Cl^-\} = \gamma_{Cl}[Cl^-] = 0.80 \cdot 0.03 = 24mM$$

Notice the significant reduction in the activity of the trivalent Fe^{3+} compared to the less reduction in the activity of the monovalent Cl^-.

4.3.3 Alkalinity and buffers

Adding acid or base to a water sample ought to change the pH of the sample accordingly. That is if the sample does not contain compounds that can react with the acid or base added either by binding or releasing protons. Compounds that can bind or release protons according to addition of acid or base, and thus decrease the actual pH change are called pH buffers. Buffers generally have the ability to maintain stable conditions within a system regardless of various changes within the system. All equilibrium reactions involving protons (H^+) or hydroxyl ions (OH^-) will have some buffer capacity against pH changes.

Phosphate is an example of a well-known pH-buffer having more than one buffer reaction:

$$H_3PO_4 \rightleftharpoons H_2PO_4^- + H^+ \quad K = 10^{-2.1} \text{ at } 25°C \tag{4.62}$$

$$H_2PO_4^- \rightleftharpoons HPO_4^{2-} + H^+ \quad K = 10^{-7.2} \text{ at } 25°C \tag{4.63}$$

$$HPO_4^{2-} \rightleftharpoons PO_4^{3-} + H^+ \quad K = 10^{-12.3} \text{ at } 25°C \tag{4.64}$$

At the point where the concentrations of, for example, the two compounds $H_2PO_4^-$ and HPO_4^{2-} are equal, the buffer capacity of equation 4.62 either binding or releasing protons will be highest.

Using the definition of the equilibrium constant K:

$$K = \frac{\{HPO_4^{2-}\}\{H^+\}}{\{H_2PO_4^-\}} = 10^{-7.2} \tag{4.65}$$

and applying equal concentrations of HPO^{2-}_4 and $H_2PO_4^-$ one obtains:

$$K = \{H^+\} = 10^{-7.2} \tag{4.66}$$

$$pH = -log\{H^+\} = -log(K) = 7.2 \tag{4.67}$$

Thus, at pH 7.2 reaction 4.63 has the highest buffer capacity. At a lower pH, say 5.2, the ratio of $\{HPO_4^{2-}\}$ to $\{H_2PO_4^-\}$ is only 0.01, meaning that 99% exists on the form $H_2PO_4^-$. At higher pH the situation is reversed. At pH 9.2, 99% exists on the form HPO_4^{2-}. In any case the equilibrium is pushed so far to either side that the buffer capacity is lost.

Alkalinity and acidity are other words for buffer capacity. Alkalinity is the capacity of a solution to neutralise strong acids. Acidity is the capacity of a solution to neutralise strong bases.

The definition of buffer capacity (β) is the amount (moles per litre) of a strong base (C_B) causing pH to increase one unit when added to a solution. The buffer capacity can also be described in terms of added strong acid (C_A) causing a decrease in pH.

$$\beta = \frac{d\,C_B}{d\,pH} = -\frac{d\,C_A}{d\,pH} \qquad (4.68)$$

If a solution contains more than one buffer, the buffer capacities for each buffer can be added. Keep in mind that water itself is a buffer according to the above definition.

$$\beta = \beta_{H_2O} + \beta_1 + \beta_2 + \ldots + \beta_n \qquad (4.69)$$

The capacity will be a function of pH and is often given as either single values at a certain pH or graphically as capacity versus pH. The buffer capacity of water [*Snoeyink and Jenkins, 1980*] can be calculated as:

$$\beta_{H_2O} = 2.3\left([H^+]+[OH^-]\right) = 2.3\left([H^+]+\frac{K_w}{[H^+]}\right) \qquad (4.70)$$

where K_w is the ionisation product constant for water. As stated above, the buffer capacity for a certain buffer has its maximum at the equilibrium constant for the reaction where the buffer gains or loses a proton. This is, however, not the case with water. Water itself is abundant in an aqueous solution whereas other buffers are dissolved in limited amounts. The buffer capacity for water B is actually lowest at pH equal to K_w whereas it increases as pH is lowered or raised.

The buffer capacity of a solution can be calculated if all compounds and equilibria in the solution are known. The buffer capacity of a simple system with one equilibrium including protons:

$$AH \rightleftharpoons A^- + H^+ \qquad (4.71)$$

where A is the compound gaining the proton, can be calculated as:

$$\beta_{HA \rightleftharpoons A^- + H^+} = 2.3\left(\frac{[HA][A]}{[HA]+[A]}\right) \qquad (4.72)$$

The calculation does not account for the buffer capacity of water, but the two values can be added in order to obtain the total capacity.

As the number of interacting proton equilibria increases so does the complexity of the calculations. The buffer capacity of a system of two interacting equilibria can be calculated as:

$$\beta = 2.3\left(C_t K_1 [H^+] \frac{[H^+]^2 + 4K_2[H^+] + K_1K_2}{\left([H^+]^2 + K_1[H^+] + K_1K_2\right)} \right) \tag{4.73}$$

where C_t is the total concentration of the three compounds and K_1 and K_2 are the respective equilibrium constants. This equation does not include the buffer capacity of water either. Keep in mind that activity coefficients may apply and that the equilibrium constants K_1, K_2 may depend on temperature and pressure.

If the pK value of interacting proton equilibria are more than two units apart, the calculation can be done for each equilibrium, adding the resulting buffer capacities [*Snoeyink and Jenkins, 1980*] as if they did not interact. The pK values for the three phosphate reactions are separated by at least two units, and the buffer capacity can thus be calculated for each reaction separately, followed by a summation with the buffer capacity of water to give the total capacity in the solution.

The buffer capacity can also be determined experimentally by adding an increasing amount (V_B) of base to a solution and recording the pH. The slope of the titration curve (dpH/dV_B) will be proportional to the buffer capacity:

$$\beta = \frac{1}{X \cdot \frac{d\,pH}{d\,V_B}} \tag{4.74}$$

where X is the ratio of base concentration to sample volume:

$$X = \frac{sample\ volume}{base\ concentration\ [OH^-]} \tag{4.75}$$

4.3.4 The carbonate system

One of the most important buffer systems in fresh water is the carbonate system, which includes at least the following two buffer equilibria (see Table 4.7 for temperature dependence of equilibrium constants):

$$H_2CO_3 \rightleftharpoons H^+ + HCO_3^- \qquad pK_1 = 6.35 \text{ at } 25°C \tag{4.76}$$

$$HCO_3^- \rightleftharpoons H^+ + CO_3^{2-} \qquad pK_2 = 10.33 \text{ at } 25°C \tag{4.77}$$

Table 4.7 Temperature dependence of carbonate equilibrium constants [Snoeyink and Jenkins, 1980]

Reaction	Temperature [°C]						
	5	10	15	20	25	40	60
$CO_{2(g)} + H_2O \rightleftharpoons H_2CO_{3(aq)}$	1.20	1.27	1.34	1.41	1.47	1.64	1.80
$H_2CO_3 \rightleftharpoons H^+ + HCO_3^-$	6.52	6.46	6.42	6.38	6.35	6.30	6.30
$HCO_3^- \rightleftharpoons H^+ + CO_3^{2-}$	10.56	10.49	10.43	10.38	10.33	10.22	10.14
$Ca^{2+} + CO_3^{2-} \rightleftharpoons CaCO_{3(s)}$	8.09	8.15	8.22	8.28	8.34	8.51	8.74

The carbonate system is influenced by carbon dioxide gas through:

$$CO_{2(g)} + H_2O \rightarrow H_2CO_{3(aq)} \quad pK = 1.47 \text{ at } 25°C \tag{4.78}$$

which depending on whether or not the solution is in contact with a carbon dioxide containing atmosphere, may alter the behaviour of the carbonate system.

Depending on the content of the solution, the carbonate system may also include precipitates like calcium carbonate, which too provides a possible exchange of carbonate:

$$Ca^{2+} + CO_3^{2-} \rightleftharpoons CaCO_{3(s)} \quad pK = 8.34 \text{ at } 25°C \tag{4.79}$$

The distribution among the three interacting dissolved compounds (CO_3^{2-}, HCO_2^- and H_2CO_3) as depending on pH can be represented in a pH-pC diagram. In such a diagram the concentration (C) of each compound is drawn as the negative logarithm (-log(C)) versus pH. Figure 4.8 shows the pH-pC diagram for a closed carbonate system with a total concentration of all three carbonate compounds of 0.1 mM.

According to the above, the closed carbonate system has maximum buffer capacities at pH 6.35 and pH 10.33, separated by at least two pH units. The buffer capacity of the two carbonate buffer reactions can thus be calculated separately along with the buffer capacity of water and summarised. Figure 4.8 shows the calculated buffer capacity for the closed carbonate system.

In open systems the buffer capacity of the carbonate system will include the equilibrium of dissolving carbon dioxide (equation 4.78) from the atmosphere, which can be assumed to contain an unlimited amount of carbon dioxide.

As long as H_2CO_3 is the major carbonate compound in the solution, the carbonate concentration will be constant as determined by the concentration of carbon dioxide in the atmosphere through (equation 4.78).

As the pH increases and the H_2CO_3 is reduced to HCO_3^- and CO_3^{2-}, the concentration of H_2CO_3 should decrease. Nevertheless, since the equilibrium between the atmospheric carbon dioxide (assumed to be abundant) and the H_2CO_3 is fixed, the concentration of H_2CO_3 cannot change. The conversion of H_2CO_3 to HCO_3^- and further to CO_3^- will not be affected by this. Thus, the total concentration of carbonate consequently increases with increasing pH.

Figure 4.9 shows the pH-pC diagram and the buffer capacity of an open carbonate system in equilibrium with atmospheric air. Notice the increasing total concentration of carbonate, compared to the constant total concentration within the closed system (Figure 4.8).

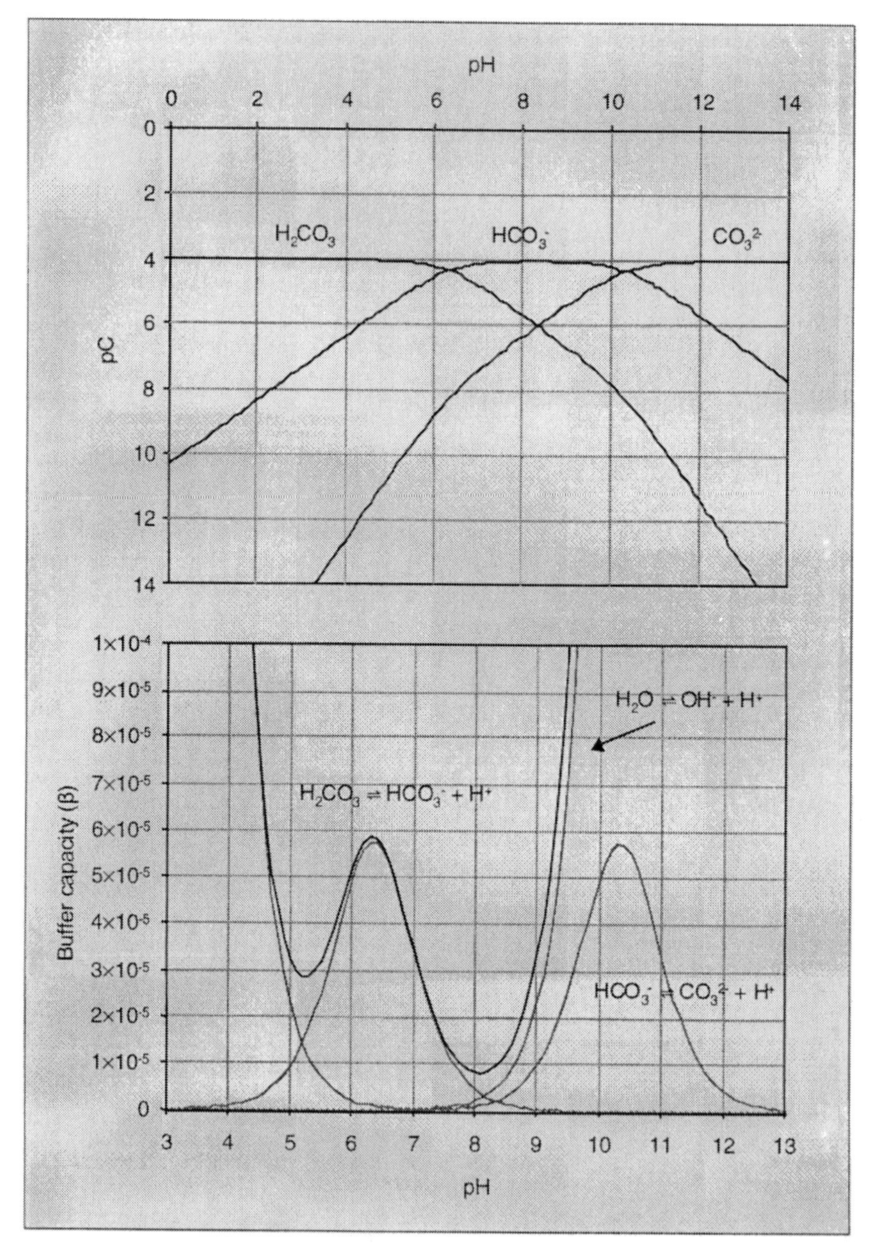

Figure 4.8 The pH-pC diagram (above) and the buffer capacity as depending on pH (below) of a closed carbonate system with a total carbonate concentration of 0.1 mM

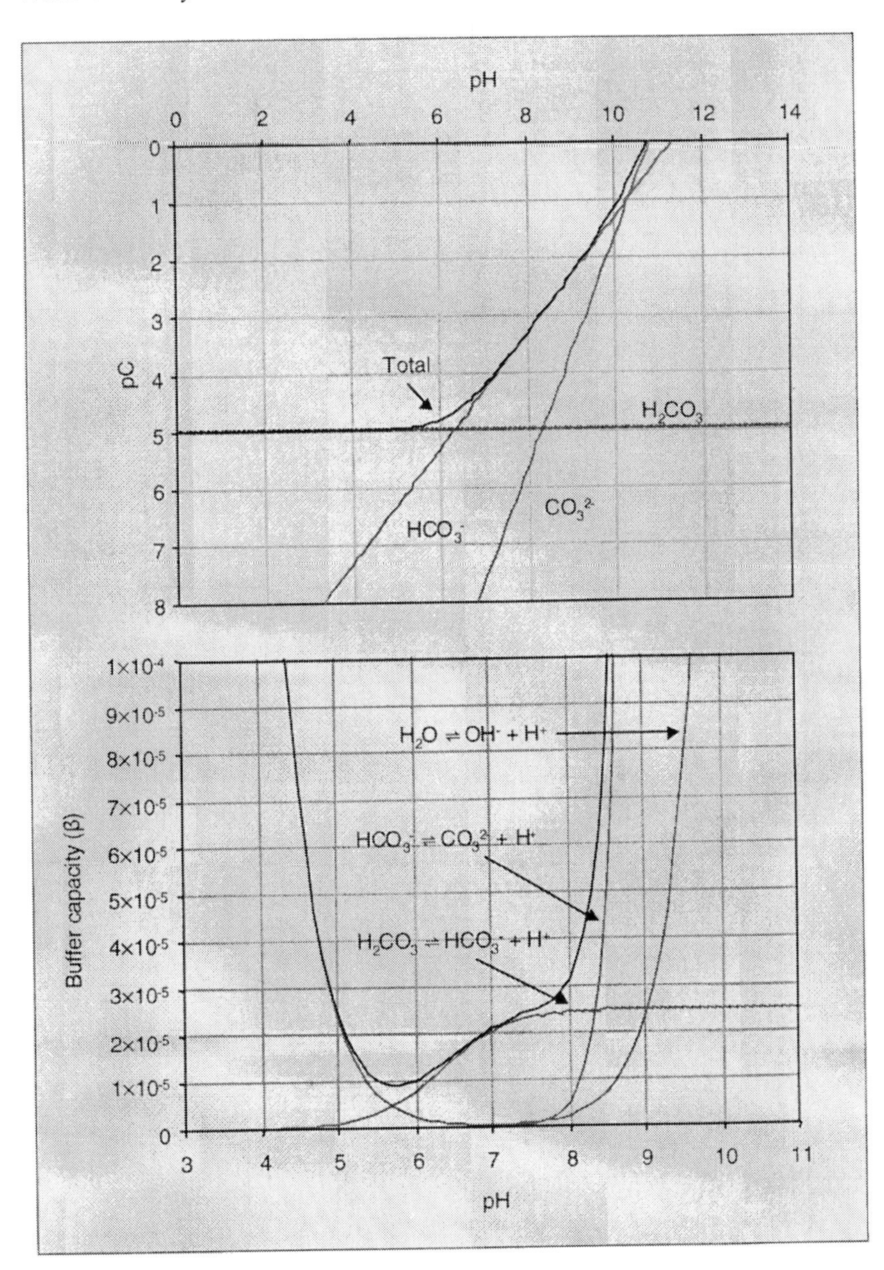

Figure 4.9 The pH-pC diagram (above) and the buffer capacity as depending on pH (below) for an open carbonate system in equilibrium with atmospheric air.

As an example of the usefulness of buffer capacity diagrams like the ones shown in Figure 4.8 and Figure 4.9, consider having to change the pH of one cubic meter of water (1000 litres) from 7 to 9. Using a (1 mol/l or 1M) solution of sodium hydroxide (NaOH), which is a strong base, how much should be added? The answer will depend on the buffer capacity of the water.

Rearranging the original definition of buffer capacity (equation) one may obtain an expression of the base required to change pH as depending on the buffer capacity:

$$C_{Base} = \int_{pH_{start}}^{pH_{end}} \beta \, dpH$$

The base needed to change pH from one value to another may thus be found by integrating the curve for buffer capacity between the two pH values. In ordinary terms, this means to find the area underneath the curve, limited by the two pH valves. For pure water the curve for buffer capacity looks like:

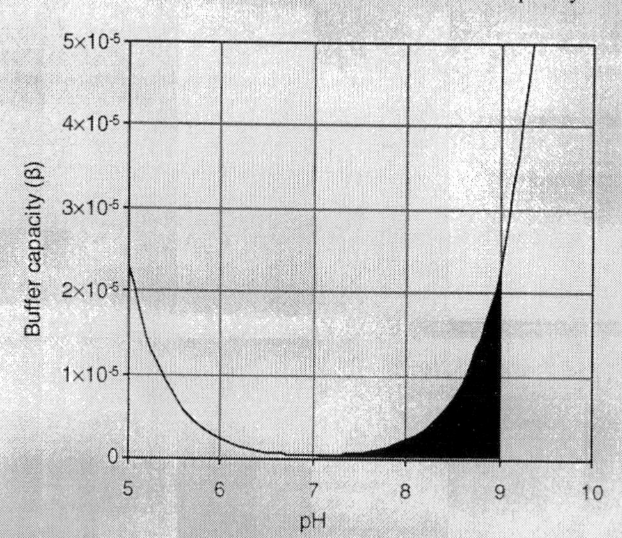

The hatched area represents the concentration of base necessary to change the pH in pure water from 7 to 9, which is about 10 μmol/l or 10 mmol in total. Using the 1M NaOH, 10 ml should thus be added.

If the water in need of the pH adjustment were to be in contact with the atmosphere, carbon dioxide would dissolve in the water increasing the buffer capacity. In that case the curve would look like:

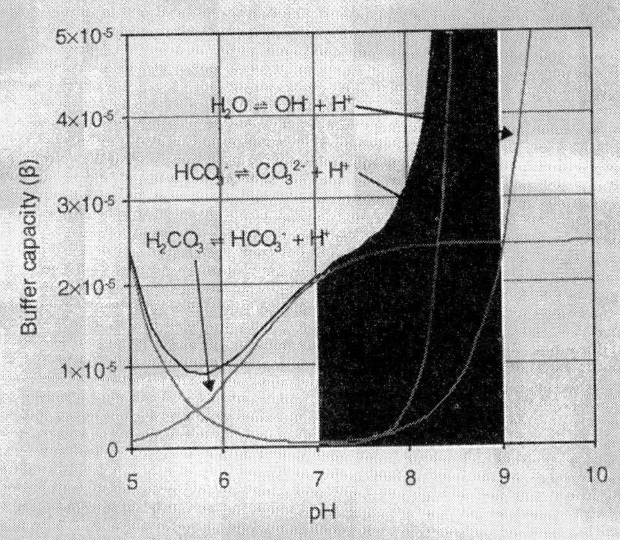

In this case, the area is significantly larger, about 160 µmol/l or 160 mmol in total. Using the 1M NaOH, 160 ml should be added instead of the 10 ml in the case of pure water.

This effect is the reason why it takes much more base to increase pH in an open system compared to a closed one.

In the example, the water was assumed completely pure with or without contact to the atmosphere. Had the water contained other compounds with buffer properties (e.g. phosphates), it might have required yet more base to change the pH.

4.3.5 Hardness

Hardness of water is an old term that relates to the soap binding and precipitating properties of the water. Earth ions, like calcium and magnesium, in the water replace sodium and potassium ions in the soap resulting in precipitation of the soap. Hardness also causes problems of scale (insoluble calcium carbonate precipitates) formation in boilers, kettles, pipes, etc. Water with a low content of calcium and magnesium is called soft, whereas water with a high content of the two ions is called hard.

The main source for hardness is calcium hydrogen carbonate or calcium bicarbonate ($Ca(HCO_3)_2$), formed within groundwater regions of limestone and chalk.

Hardness is divided into different subterms:

Total hardness
The total concentration of calcium and magnesium ions in a water sample.

$$Total\ hardness\ (mg/l) = [Ca^{2+}]_{(mg/l)} + [Mg^{2+}]_{(mg/l)} \tag{4.80}$$

Theoretically, other divalent ions like iron (Fe^{2+}) and manganese (Mn^{2+}) also contribute to the hardness of water. However, the concentration of these two ions in natural waters is usually much lower than the concentration of calcium and magnesium.

Hardness is often given in terms of the degree of hardness (°dH):

$$1°dH = 0.179\ mmol\ Ca/l = 7.13\ mg\ Ca/l \tag{4.81}$$

$$1°dH = 0.179\ mmol\ Mg/l = 4.35\ mg\ Mg/l \tag{4.82}$$

$$Total\ hardness\ (°dH) = \frac{[Ca^{2+}]_{(mg/l)}}{7.13} + \frac{[Mg^{2+}]_{(mg/l)}}{4.35} \tag{4.83}$$

Total hardness of tap water may vary from 0 (very soft water) to about 30 (hard water). In some cases the hardness might even exceed 30. The total hardness can theoretically be divided into temporary and permanent hardness.

Temporary hardness
Also called carbonate hardness or bicarbonate hardness. The part of the total hardness that is related to carbonates and bicarbonates of calcium and magnesium (e.g. $Ca(HCO_3)_2$). Temporary hardness can be removed by boiling during which the carbonate is released as carbon dioxide

$$Ca(HCO_3)_{2(aq)} \rightleftharpoons CaCO_{3(s)} + H_2O + CO_{2(g)} \tag{4.84}$$

The temporary hardness can be calculated on the basis of the bicarbonate concentration:

$$Temporary\ hardness\ (^\circ dH) = \frac{[HCO_3^-]_{(mg/l)}}{2 \cdot 7.13} \cdot \frac{40}{60} \tag{4.85}$$

Due to the definition, the calculated temporary hardness may exceed the total hardness. If the water sample contains sodium or potassium carbonates, these will too contribute to the measured temporary hardness. Sodium and potassium, however, is not included as hardness because they do not form insoluble complexes. Thus the temporary hardness is not allowed to exceed the total hardness.

Permanent hardness

The difference between the total hardness and the temporary hardness is the permanent hardness, which is due to the calcium and magnesium ions corresponding to anions like sulphate, phosphate, fluoride, etc that are present in tap water in some areas. The permanent hardness cannot be removed by boiling but requires ion exchange or reverse osmosis.

Calcium hardness

In most cases calcium is the primary source of hardness. Therefore, hardness is sometimes given as calcium hardness, which is the part of the total hardness that is related to calcium.

Water softening

The problems with hard water concentrate on the precipitation and complexion of calcium and magnesium ions at places where the precipitate is unwanted (e.g. scales in boilers and pipes). Hard water is therefore often softened in order to eliminate or minimise these problems. Water softening is the removal of calcium and magnesium ions. The process can be one of the following:

- Precipitation with calcium hydroxide or lime ($Ca(OH)_2$):

$$Ca(OH)_{2(aq)} + Ca(HCO_3)_{2(aq)} \rightleftharpoons 2CaCO_{3(s)} + 2H_2O \tag{4.86}$$

  The process is also called the Clark process or clarkening. It only removes the temporary hardness, not the permanent one.

- Precipitation with sodium carbonate or soda (Na_2CO_3). The process removes both temporary and permanent hardness.

- Ion exchange. Removes both permanent and temporary hardness by exchanging calcium and magnesium ions with sodium or potassium that do not precipitate.

- Complex binding with e.g. polyphosphate ($P_6O_{18}^{6-}$).

4.4 Water analysis

Since the properties of water and the components it may contain, influence the behaviour and application range of the water, it is necessary to keep track of the quality by performing various analyses and tests. Water analysis covers the determination of more or less any compound or group of compounds that may be contained within water and the field is thus very widespread. For the purpose of this section, water analysis will, however, be limited to the area of potential interest for tap water hydraulic systems including areas of importance to performance, durability, and applications of the systems.

Water analysis can be divided into the same groups as the compounds that may be found within the water: organic and inorganic. Inorganic analysis includes the determination of different ions and gases whereas organic analysis covers the area of organic molecules. Besides the determination of specific compounds, water analysis also provides a range of bulk or collective analyses that can be used to obtain overall values for water quality.

4.4.1 Bulk analysis

One of the most common parameters to look for in water is the amount of dry matter. The method is quite simple. A known volume of water is dried at 105°C until it does not change weight any more, that is when all the water has evaporated. The remains are weighed and reported as dry matter usually in mg/l. The method works best on heavily soiled samples with a high content of dry matter, but it may also be applied to more pure systems. The accuracy will, however, be limited to the accuracy of the weight measurement and thus larger samples are required for lower content of dry matter.

The substances contained within the dry matter include all dissolved and suspended solids within the original sample (most gases are removed during drying) such as salts, minerals, organic substances, etc. Further information on the substances can be obtained by heating the dry matter further to about 550°C. At this high temperature the organic matter within the sample is combusted to carbon dioxide, leaving only inorganic salts and minerals. Weighing the sample again yields the amount of inorganic dry matter whereas the difference between the two dry weights is the amount of organic dry matter. For relatively pure water the concentration of organic compounds is usually low, thus salts and minerals primarily constitute the dry matter.

The analysis of dry matter may, in some cases, be disturbed by the formation of crystal water. When the sample is dried at 105°C, some water may be bound within the crystallising salts. When the sample is further heated, the crystal water evaporates, imitating organic matter being combusted. The risk of formation of crystal water increases with increasing content of inorganic matter.

The approximate amount of dissociated salts (ions) within a water sample may also be estimated by measuring the conductivity of the water sample (see sections 4.3.2 and 4.4.4).

4.4.2 Inorganic compounds

There exists a large number of methods to analyse inorganic compounds in water; practically one for each possible ionic species. Table 4.8 shows a selection of inorganic parameters that may be measured in water samples. Some methods use titration, some use development of colour followed by spectrophotometry, and some use other techniques.

This section will not describe the specific methods, but lists some of the inorganic compounds and their relevance to tap water hydraulics.

Calcium (Ca^{2+}) and Magnesium (Mg^{2+})
Major contributors to formation of scale and other hard precipitates. Calcium and magnesium are, together with carbonate, the basis for water hardness (see section 4.3.5).

Chloride (Cl^-) and Fluoride (F^-)
These provide a potential threat towards localised corrosion of passive metals such as stainless steels.

Table 4.8 Examples of inorganic parameters to be analysed in water samples

Cations	Anions	Heavy metals	Gases
Sodium (Na^+) Potassium (K^+) Ammonia (NH_4^+) Iron (Fe^{2+}/ Fe^{3+}) Manganese (Mn^{2+}) Calcium (Ca^{2+}) Magnesium (Mg^{2+})	Chloride (Cl^-) Nitrate (NO_3^-) Sulphate (SO_4^{2-}) Nitrite (NO_2^-) Phosphate (PO_4^{3-}) Fluoride (F^-)	Copper (Cu^{2+}/Cu^+) Zinc (Zn^{2+}) Lead (Pb^{2+}) Mercury (Hg^{2+}/Hg_2^{2+}) Chromium (Cr^{2+}/Cr^{3+})	Hydrogen sulphide (H_2S) Carbon dioxide (CO_2) Oxygen (O_2)

Manganese (Mn^{2+})
Can cause chemical/microbiological deposition of manganese dioxide, which may cause localised corrosion of stainless steels. Although it is relatively harmless, large amounts of manganese dioxide turn the water brown and make it look unhealthy.

Ammonia (NH_4), Nitrate (NO_3), and Nitrite (NO_2)
Can be oxidised/reduced microbiologically to nitrogen gas, forming gas pockets in high sections of the hydraulic system. Nitrite is somewhat toxic. The nitrogen compounds may serve as an energy source for microbial growth (see the chapter on microbiology)

Sulphate (SO_4) and Hydrogen sulphide (H_2S)
Hydrogen sulphide is corrosive to both passive and active metals and is highly toxic to humans. Sulphate is not corrosive but may be reduced microbiologically to sulphide, which is highly corrosive. In most systems, the reduction of sulphate to sulphide may only be performed by microorganisms and sulphide can thus be used as a fingerprint for microbial growth. Due to its corrosive behaviour, sulphide may also be used as an indicator for microbiologically influenced corrosion.

Phosphate (PO_4)
May provide a buffer against pH changes (see section 4.3.3). Phosphate will also provide a source of phosphor required for growth of microorganisms. If the microbial growth is phosphor limited (see chapter on microbiology) adding phosphate may cause increased microbial growth.

4.4.3 Organic compounds

The content of organic compounds within a water sample is often measured as one parameter. The organic compounds may be numerous and thus nearly impossible to classify using specific analyses. Instead, methods have been developed to measure the content of organic compounds as e.g. the content of organic carbon (carbon from organic compounds) or the amount of some oxidising compound needed to oxidise the organic compounds to carbon dioxide. The simplest way of assessing the organic content of a sample is combusting the sample and weighing the ashes (see section 4.4.1). However, this simple method may be difficult to interpretate in some cases. In the following, some of the most common and more reliable methods for analysing the organic content of a water sample are briefly described.

Biological Oxygen Demand (BOD)
The method is based on the fact that aerobic microorganisms use oxygen when degrading organic matter. Briefly, a sample is sealed in a gas tight container and

the content of oxygen is measured after some time. The difference in oxygen content between the original sample and the stored sample is taken as a measure for thc organic content of the sample. Usually the sample is stored for 5 or 7 days and the results are given in mgO_2/l as BOD_5 or BOD_7. Biological oxygen demand has been used extensively in many fields. However, due to the time scale and the accuracy associated with the method, it has been replaced by other methods. The BOD method is, furthermore, limited by the fact that not all organic compounds are available for utilisation by the aerobic bacteria.

Chemical Oxygen Demand (COD)
As the name indicates, this method uses a chemical reaction instead of a biological one. The time needed for the analysis is thus reduced from several days to a few hours. The organic compounds within the sample are oxidised by potassium dichromate, a strong oxidising agent. Briefly, potassium dichromate and sulphuric acid are added to the sample and the sample is boiled for two hours, during which the following reaction occur:

$$3CH_2O + 2K_2Cr_2O_7 + 16H^+ \rightleftharpoons 3CO_2 + 4Cr^{3+} + 4K^+ + 11H_2O \qquad (4.87)$$

where CH_2O represents the organic compound. Hereafter, the remainder of dichromate within the sample is measured by means of titration or spectrophotometry. The spent dichromate is given as the equivalent oxygen amount to oxidise the same amount of organic matter.

$$CH_2O + O_2 \rightleftharpoons CO_2 + H_2O \qquad (4.88)$$

Thus, 2 mol (294g) potassium dichromate equals 3 mol (96g) oxygen or 1 gram oxygen equals 3 grammes potassium dichromate.

The method may be disturbed if the sample contains reduced inorganic substances. In that case some of the dichromate will be used to oxidise the reduced compounds resulting in overestimation of the organic content.

Potassium permanganate
Potassium permanganate is another oxidising agent, commonly used in measurements of organic matter. The method is similar to COD but since permanganate is a weaker oxidiser compared to dichromate the oxidisation will be less complete. Potassium permanganate and sodium hydroxide are added to the sample, which is boiled for 10 minutes. The reaction is as follows:

$$5CH_2O + 4KMnO_4 \rightleftharpoons 5CO_2 + 4Mn^{2+} + 4K^+ + 5H_2O \qquad (4.89)$$

As with COD, the result is given as equivalent oxygen. 4 g potassium permanganate equals 1 g oxygen.

Total Organic Carbon (TOC)
This method also uses a chemical oxidising agent at a high temperature, but instead of measuring the amount of oxidiser used, it measures the amount of carbon dioxide produced. The result is given as total organic carbon.

The method is fast (few minutes) and can be used to detect organic matter in quite low concentrations (1mg/l). It has, however, some limitations in terms of contamination of the analysis apparatus by suspended solids.

4.4.4 The pH electrode and other online sensors

The pH electrode
The pH of an aqueous solution can be measured continuously as a potential difference between a reference electrode and a hydrogen glass electrode (Figure 4.10). The glass electrode consists of a bulb of thin glass that is sensitive to changes in the proton activity $\{H^+\}$. The glass electrode is filled internally with a solution of known pH; usually a 0.1 M HCl solution is used. An internal reference electrode, usually a silver wire coated with silver chloride (also called an Ag/AgCl electrode) is placed in the internal solution.

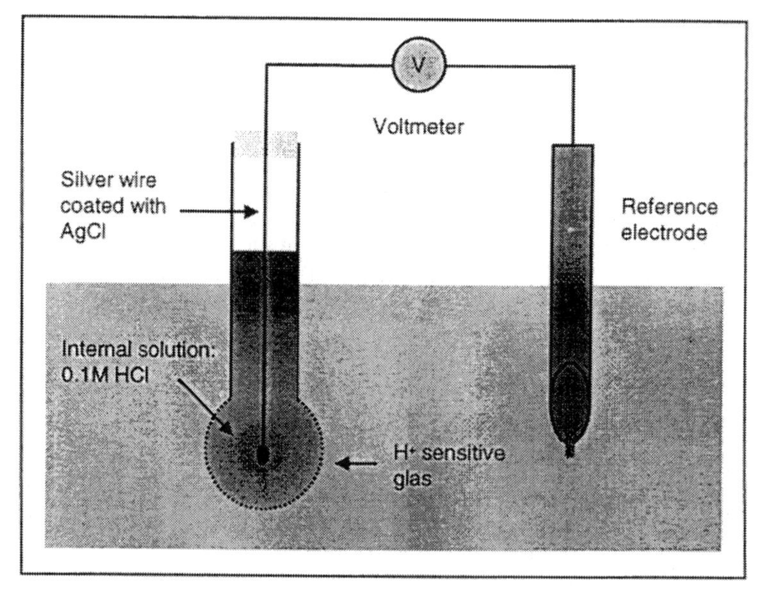

Figure 4.10 The two-electrode set-up used to measure pH

Due to the difference in proton activity between the internal and the external solution a potential develops across the proton sensitive glass membrane. In order to measure this potential, a reference electrode with a stable potential is needed. Normally, a calomel electrode is used as the reference. The potential measured between the glass electrode and the reference can be related to the pH of the solution through:

$$pH = constant - \frac{E_{measured}}{constant} \tag{4.90}$$

For practical purposes the two electrodes - the glass electrode and the reference electrode - are usually combined in a single electrode. The pH electrode is connected to a pH meter, which is basically a high resistance voltmeter with the scale calibrated to read pH instead of volts.

Before performing a pH measurement, the pH meter has to be calibrated in order to determine the two constants in the above equation. In order to calibrate a pH meter, at least two different solutions of known pH must be used. Usually solutions buffered at pH 7.00 and pH 4.01 are used as references. Adjusting both constants (origin and sensitivity), the pH meter is brought to read the correct value in the two reference solutions and is then ready for measurements.

The constants in equation 4.90 are temperature sensitive. Therefore, the temperature of reference solutions at the time of calibration should be the same as the temperature of the unknown sample.

The conductivity electrode

Measurements of conductivity can be used as an approximate estimate of the concentrations of ions in solution (see section 4.3.2). Table 4.9 shows the conductivity of solutions with different potassium chloride concentrations. Remember that besides the ion concentration, the conductivity also depends on the charge of the ions.

Table 4.9 Conductivity (mS/cm) of a solution with different KCl concentration (mol/l) at 25°C

[KCl]	Conductivity	[KCl]	Conductivity
0.0001	1.494	0.05	666.8
0.001	14.70	0.1	1,290
0.005	71.78	1	11,190
0.01	141.3		

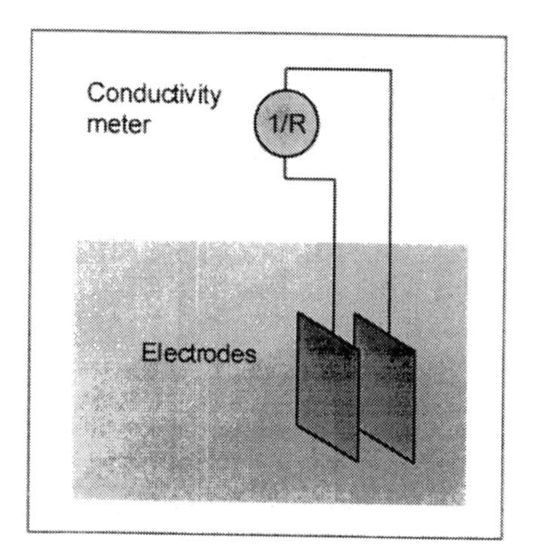

Figure 4.11 The principle of measuring conductivity of a water sample

The measurement of conductivity is a simple electronic measurement of the resistance (the reciprocal conductivity) between two electrodes immersed in a water sample (see Figure 4.12). The distance between, and orientation of, the two electrodes has to be fixed and correlated for, since the measured resistance also depends on these parameters. For practical reasons the two electrodes are fixed within one electrode, the conductivity electrode.

The conductivity measured in a solution is given as:

$$\kappa = \frac{k_{electrode}}{R} \tag{4.91}$$

where κ is the conductivity in e.g. mS/cm or $M\Omega^{-1}cm$, k is a temperature dependent constant related to the electrode given as cm^{-1} and R is the resistance measured in e.g. $M\Omega$.

For calibration of the conductivity meter and electrode, in order to determine the constant k in the above equation, a known solution of 0.01 M potassium chloride is usually used. Due to the temperature dependence (see Table 4.10) it is essential that the conductivity electrode and meter are calibrated at a known temperature usually 25°C. Some conductivity meters are equipped with automatic temperature compensation enabling them to be used at a range of temperatures. However, because the temperature dependence is not necessarily the same for different ions it is recommended to perform the conductivity measurements at about the same temperature as the temperature of calibration.

Table 4.10 Conductivity of a 0.01 M KCl solution at different temperatures

Temperature (°C)	Conductivity (mS/cm)
5	217.6
10	192.2
15	171.0
20	155.4
25	141.3
30	130.0

The conductivity range of a certain meter depends mainly on the electrode. Different electrodes are commercially available. Some have quite a narrow conductivity range whereas others may be used over a broader range.

The oxygen sensor

The construction of the oxygen electrode (see Figure 4.12) is somewhat similar to the pH electrode, but it uses a current measurement usually along with an applied potential instead.

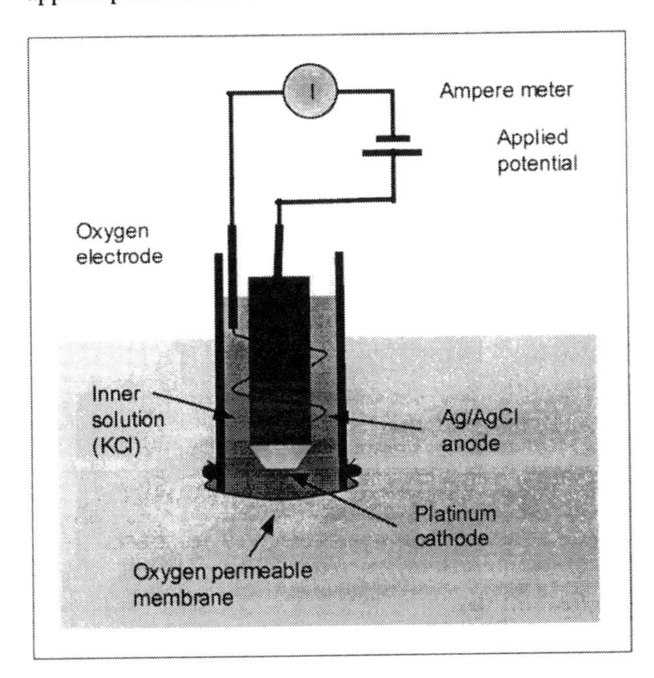

Figure 4.12 Schematic representation of an oxygen electrode

The oxygen electrode is based on the electrochemical reduction of oxygen:

$$O_2 + 2H^+ + 2e^- \rightleftharpoons H_2O_2 \quad (4.92)$$

$$H_2O_2 + 2H^+ + 2e^- \rightleftharpoons 2H_2O \quad (4.93)$$

Where e^- represents electrons that are consumed at the cathode (an electrode removing electrons from the solution). One oxygen molecule (O^2) requires four electrons to be reduced by the cathode. The anode (an electron supplying electrons to the solution) returns the electrons to some compound in the solution again.

A common anode used for oxygen electrodes is the silver/silver chloride (Ag/AgCl) electrode supplying the following anodic reaction:

$$Ag_{(s)} + Cl^- \rightleftharpoons AgCl_{(s)} + e^- \quad (4.94)$$

Using the Ag/AgCl electrode requires the internal solution in the electrode to contain chloride. Usually, a half-saturated potassium chloride (KCl) solution is used for oxygen electrodes with Ag/AgCl anodes.

In order to drive, the cathodic reduction of oxygen and the anodic oxidation of silver chloride, a potential of around 0.8 V is applied between the cathode and the anode (see Figure 4.12). Applying the potential causes a reduction of practically all the oxygen within the internal solution of the oxygen sensor.

The final, and the most important, part of the oxygen sensor is a membrane, separating the internal solution from the test solution. The membrane is semi-permeable for oxygen, meaning that only oxygen can penetrate it and thus be transferred from the test sample to the internal solution. All ionic and particulate matter is kept from entering the internal solution and contaminating the sensor.

Since the applied potential cause reduction of all the oxygen within the internal solution, the rate of which oxygen is reduced or the rate of which electrons are produced at the cathode, is proportional to the diffusion rate of oxygen through the semi-permeable membrane. Furthermore, since the oxygen concentration in the internal solution is practically zero or at least constant, the diffusion rate of oxygen across the membrane only depends on the oxygen concentration within the solution of interest.

Now, measuring the rate of electron production at the cathode, which is equal to the current that flows between the cathode and the anode, completes the actual measurement (see Figure 4.12).

The oxygen electrode is calibrated using two points: one measurement in air saturated with water and one measurement in water without oxygen. Atmospheric air saturated with water has a certain content of oxygen depending on the temperature and pressure, and can thus be calculated in each case.

References

Christenson, P. G. and Gieskes, J. M., *Journal of Chemical and Engineering Data* **16**, 398 (1971).

Culberson, C. and Pytkowicz, R. M., *Marine Chemistry* **1**, 309 (1973).

Granat, L., *Tellus* **24**, 550-556 (1972).

Harned and Owen, "*The Physical Chemistry of Electrolytic Solutions*," van Nostrand Reinhold, New York, 1958.

Hill, H.M., "*Composition of Seawater, Comparative and Descriptive Oceanography*," Wiley-Interscience, New York, 1963.

Russel, L. L., "*Chemical of Groundwater Recharge with Wastewaters*", Ph.D. Thesis, University of California, Berkeley, 1976.

Snoeyink and Jenkins, "*Water Chemistry*," John Wiley & Sons, Inc., New York, 1980.

Stumm, W. and Morgan, J. J., "*Aquatic Chemistry*", John Wiley & Sons, Inc., New York, 1981.

Whitfield, M., "Self-Ionization of Water in Dilute Sodium Chloride Solutions from 5-35°C and 1-2000 Bars," *Journal of Chemical and Engineering Data* **17**(2), 124 (1972).

5 Water microbiology

Bo Højris Olesen and Bo Frølund
Danish Technological Institute

Particularly with tap water hydraulic systems in mind this chapter takes the reader through basic terms of microbiology, discussions of microbial growth, and possible actions of prevention and mitigation of problems caused by microorganisms. A description of general analytical methods in microbiology is also included, with a view to assisting the reader in understanding the test results.

5.1 Introduction to microbiology

Microbiology is per definition the study of living organisms that are too small to be seen without optical aids. Microorganisms, the common denominator for these small creatures, cover a very large and highly diverse group, including e.g. bacteria, algae and yeast. The size of microorganisms ranges from below 0.1 µm and upward (1000 µm equals 1 mm). Viruses, being part of the microorganisms, though they are not real living organisms, range below 0.1 µm in size. Bacteria are normally between 0.1 µm and 1 µm; above 1 µm larger more complex organisms may be found. The human eye is not capable of recognising objects smaller than 100 µm or distinguishing details on objects smaller than 1 mm, so this roughly provides the upper limit of microbiology.

Whether we are aware of it or not, microorganisms are a part of our daily life. Each square centimetre of our skin is covered with thousands of these small creatures and many of our bodily functions are actually carried out by microorganisms. The human colon, for example, contains so many microorganisms, that they constitute about half of the total colon weight. Microorganisms also drive many of the external processes we depend upon, like for example the production of many dairy products and pharmaceuticals or the treatment of municipal wastewater.

Microorganisms, unfortunately, do not distinguish between processes that are useful in terms of human activity, and processes that are harmful. Like most of the higher developed plants and animals, microorganisms live by very simple rules: stay alive and reproduce. In order to do so, they consume any usable food source available, gain as much energy as possible, and excrete a variety of by-products. The presence of microorganisms and/or their by-products is often unwanted. Microorganisms or their by-products cause many diseases, attacking plants and animals (including humans). In most technical systems, like the tap water hydraulic systems addressed in this book, microorganisms and their excretions are also undesirable. The nature of problems caused by micro-organisms in such technical systems could typically be:

- Increased corrosion rate or localised corrosion
- Loss of system performance
- Increased energy consumption due to e.g. increased friction or decreased heat transfer
- Poor hygiene standard and risk of contaminating products
- Bad odour and increased health risks for workers

Due to the size of the microorganisms, they can enter practically any system. Our surroundings contain billions of microorganisms and they can easily spread through air, water, people, etc. Most microorganisms can be killed or inactivated somehow. However, the action required to kill one microorganism might not affect another. It therefore requires quite a lot of effort to construct and maintain technical systems that do not contain microorganisms (such systems are said to be sterile), and to the regret of many system designers and constructors, it often fails.

5.1.1 The cell

All living organisms, like plants and animals, are composed of small individual cells. Most of these cells cannot function if they are removed from the main organism. Microorganisms, on the contrary, are single cell organisms. They can live by themselves and do not need to be grouped together in a larger structure. Many microorganisms, though, live in small communities or in symbioses with a larger host organism (like in the human colon). However, if one of these microorganisms is isolated and given similar conditions it will be able to function alone.

All these cells, both the microorganisms and the individual cells of larger organisms, have several structural elements in common. They are all small "sacks" with something inside.

The "sack" itself provides a barrier to the surroundings, allowing some compounds to pass and blocking others.

Inside the cell lies the code for how the cell behaves along with some tools that the cell uses to function. Beyond these similarities the cells of different life forms start to deviate more or less. Animal cells, for example, are composed mostly of water, whereas plant cells are denser and structurally quite different. On the other hand, the cells within a mouse are not much different from the cells of an elephant.

The single cell organism (see Figure 5.1) consists of a layered cell wall filled with a fluid called cytoplasm. The cell wall controls what enters and leaves the cell. Nutritional compounds are transported into the cell, and different by-products are excreted to the surroundings.

The cytoplasm inside the cell contains the organelles or tools (e.g. ribosomes), that the cell uses to function along with the specific code for the particular organism, enclosed in a complex chemical structure, called Deoxyribo-Nucleic-Acid, or just DNA. In some cells the DNA is contained within a so-called nucleus. In others it floats around within the cytoplasm. The cytoplasm may also contain a variety of nutritional compounds, intermediate products and by-products like proteins, carbohydrates, salts, etc that are part of the metabolism of the cell. On the outside, some cells are equipped with one or more flagella or pili, which to some extent enables them to move and orient themselves.

A great deal of energy is required to maintain the many different structures that constitute a cell, in particular in replicating these structures when the cell divides. The cell therefore needs to continuously generate energy.

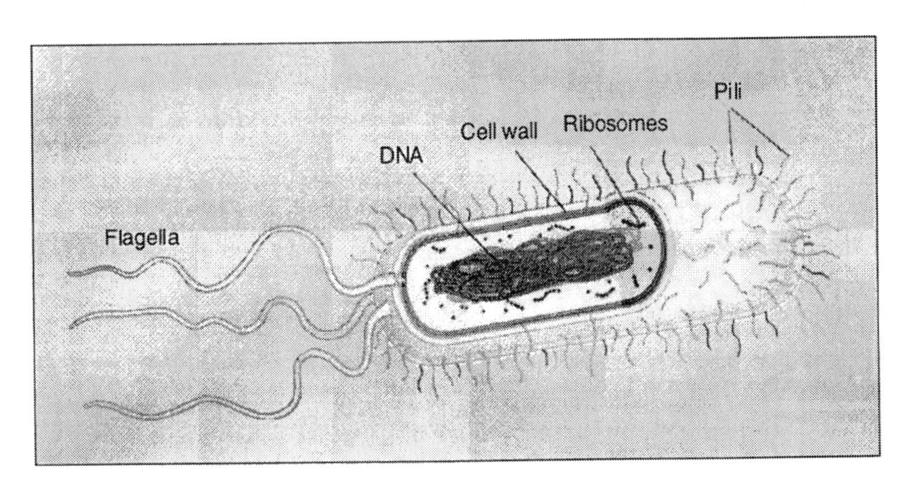

Figure 5.1 Schematic representation of a typical cell

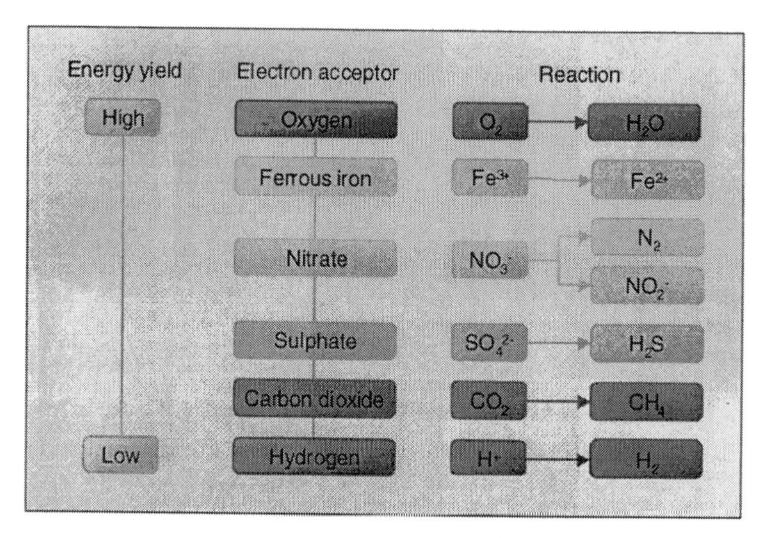

Figure 5.2 Energetic series of selected electron acceptors

Some microorganisms can use energy from light, whereas others use the energy that is released through different chemical reactions:

$$A_{Oxidised} + B_{Reduced} \Rightarrow A_{Reduced} + B_{Oxidised} + Energy \qquad (5.1)$$

In the chemical process, electrons are transferred from the compound being oxidised (B) to the compound being reduced (A). The compound donating electrons, A, (called the electron donor) is often the organic material, which the cell uses as its carbon source. The compound accepting the electrons, B, (called the electron acceptor) can be one of several inorganic compounds (see Figure 5.2). The microorganisms using chemical reactions to generate energy will always chose the electron acceptor which gives the highest possible energy yield. Nevertheless, most microorganisms are only capable of using one electron acceptor, and there will therefore be a close connection between the electron acceptor available and the microorganisms present.

The processes used to generate energy, maintaining structure, etc are controlled by a large number of complex molecules produced by the cell. The recipe or code for each of these molecules and thus the guidelines for cell behaviour is stored within the DNA.

In order to replicate itself, the cell must synthesise more than 1000 different molecules. The DNA within the cell may contain the code for more than three times as many compounds, but only the code sequences for the compounds currently in use are "turned on". The rest of the code sequences may be turned on if the conditions change and the cell, for example, needs to use another food source. In that casc it may take a while before the cell is adjusted to the new conditions and has turned the right code sequences on and off.

When a cell divides and replicates itself, the DNA code within the cell is also replicated. In copying the large DNA molecules, mistakes or errors are often made. These errors may cause the new cell to behave differently. It may for example no longer be able to generate energy, or maintain some of the important internal organelles. Therefore, most of these mutations, as the errors are called, will cause the new cell to die. In some cases, however, the new mutated cell is actually improved compared to the old one, and it may be better suited to live in the particular environment. If that is the case, a new microorganism is created. This is basically the same evolution process as the ones involved with larger organisms, only the process may happen much faster with microorganisms, as their generation time may be very short.

5.1.2 Taxonomy

It is estimated that hundreds of thousands of different microorganisms exist on the earth. However, only a very small fraction of these have been studied. The cells that have been isolated and studied have been given unique names, sometimes depending on their behaviour, shape, or the environments from which they were isolated. The organisms are given both a genus name (like a surname) and a species name (like a first name). The genus name represents a group of related species. The names are written with the genus name first followed by the species name, like for example *Saccharomyces cerevisiae* (yeast), *Dictyostelium discoideum* (mold), or *Escherichia coli* (bacterium).

In order to obtain an overview of the many different organisms, various suggestions have been put forward as to how the organisms could be grouped. One grouping, based on the physical structure of the microorganisms, is shown in Figure 5.3. This scheme divides the majority of all microorganisms into two structural types: prokaryotic and eukaryotic. Prokaryotic organisms are relatively simple in structure while eukaryotic organisms are more complex.

The division of the single cell organisms into prokaryotes and eukaryotes is largely based on one single structural difference between the two cell types. In the prokaryotic cell, the DNA floats around within the cell, whereas within the eukaryotic cells the DNA is contained within a so-called nucleus. The nucleus of the eukaryotic cell may contain several DNA molecules called chromosomes (human cells, for example, contain 23 pairs of chromosomes) whereas the

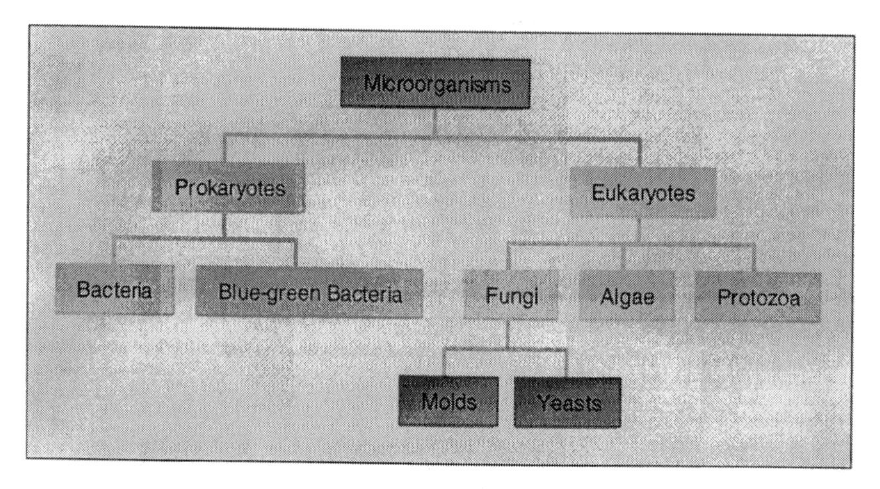

Figure 5.3 Grouping of microorganisms in terms of physical structure

prokaryotic cell contains only one. Furthermore, the eukaryotic cells may contain many different organelles used for specific processes, whereas the prokaryotic cells only hold a limited number of organelles (see Figure 5.1).

Each of these two groups are further divided. The prokaryotic group includes bacteria and blue-green bacteria, and the eukaryotes, the more complex organisms, include fungi, algae and protozoa. The subgroup, fungi, is further divided into molds and yeasts. Viruses, which are not really living organisms, but rather small bits of biological code that needs a host cell to be reproduced, are not included in the scheme.

Microorganisms can also be classified by placing them in broad physiological groups depending on their source of carbon and energy (see Figure 5.4).

Organisms using chemical reactions to generate energy are called chemotrophs whereas organisms using light as their source of energy are called phototrophs. Named by their source of carbon, organisms utilising organic matter are called heterotrophs, whereas organisms using inorganic carbon are called autotrophs. By combining the carbon and energy source, four major nutritional categories can be classified. The chemoheterotrophs, which use organic material as carbon source and chemical reactions for energy, include the majority of all bacteria. The photoautotrophs, which use light as energy source and CO_2 as carbon source, include algae and photosynthetic bacteria.

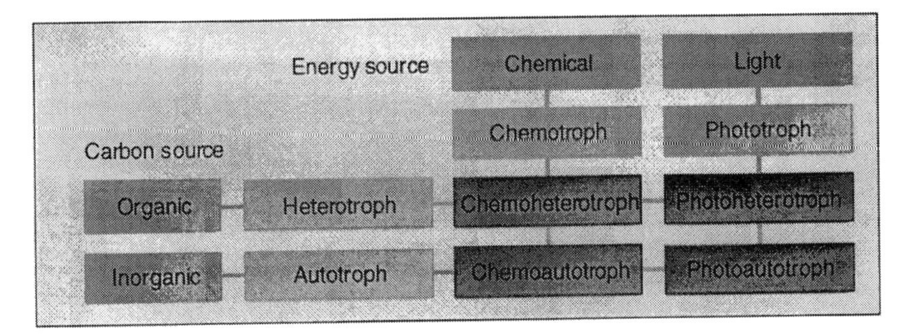

Figure 5.4 Classification of microorganisms in terms of carbon and energy source

Microorganisms, particularly the chemotrophic bacteria, may also be classified by their specific source of energy depending on which compound they use as electron acceptor in the energy providing chemical reaction. Cells using nitrate (for example) as electron acceptor are called nitrate reducers, whereas cells using sulphate are called sulphate reducers etc.

Some cells are able to use more than one compound for energy generation e.g. O_2 and NO_3. These cells are said to be facultative. In contrast, the term obligate is used to describe cells that can only use one compound e.g. obligate aerobic bacteria (bacteria that can only live in the presence of oxygen).

Bacteria can furthermore be grouped from their responses to environmental factors like temperature, hydrostatic pressure, pH and salinity. Thermophilic and hyper-thermophilic microorganisms for example can live at high temperatures (>45°C and >80°C, respectively), whereas acidophilic microorganisms can live at very low pH values (pH down to 1) and barophilic microorganisms can live at high pressure (100-1000 atm).

5.2 Microbial growth

Rather than referring to an increasing size of the single cell, the term microbial growth refers to the increasing number of cells or the increasing total mass of cells resulting from cell reproduction. The rate at which the cells divide or reproduce is called the growth rate (usually represented with the Greek letter μ).

Besides various external factors, such as temperature and pH, the growth rate of a specific type of cell may depend on the presence and concentration of all the different substrates needed for the cell to create a replica of itself. At a minimum, these compounds normally include carbon, nitrogen, phosphorus, and sulphur in some organic or inorganic form that the specific cell can utilise. Some cells may, for example, use cellulose (a quite complex organic compound) as their source of carbon, whereas other cells need less complex compounds such as e.g. glucose instead.

For each compound, required for cell reproduction, a yield coefficient (y) may be found, describing the ratio between the amount of compound utilised and the amount of cells produced:

$$y_n = \frac{gram\ of\ cells}{gram\ of\ substrate\ n} \tag{5.2}$$

The yield coefficient would also include the amount of substrate needed for the basic cell metabolism (for the cell to stay alive). Furthermore the yield coefficient will to some extent depend on the specific system. Therefore the terms y_{max} and y_{obs} are sometimes used as the maximum and the observed yield coefficient.

$$y_{obs} \leq y_{max} \tag{5.3}$$

The relationship between the nutritional compounds available and the growth rate is called kinetics. In general, the more food there is available, the faster the cells grow. However, as in many other applications, there exists a maximum for how fast the cell can divide. Depending on the cell type and the conditions at which it grows, the relationship between food and growth rate may vary. Most cells, though, follow what is called Monod kinetics (see Figure 5.5).

The mathematical basis for this kinetic behaviour, as built on experimental observations, is given as:

$$\mu = \frac{S}{S + K_s} \mu_{max} \tag{5.4}$$

where S is the substrate (food) concentration, μ is the specific growth rate, μ_{max} is the maximum growth rate, and K_s is the substrate concentration at which the specific growth rate equals half of the maximum growth rate.

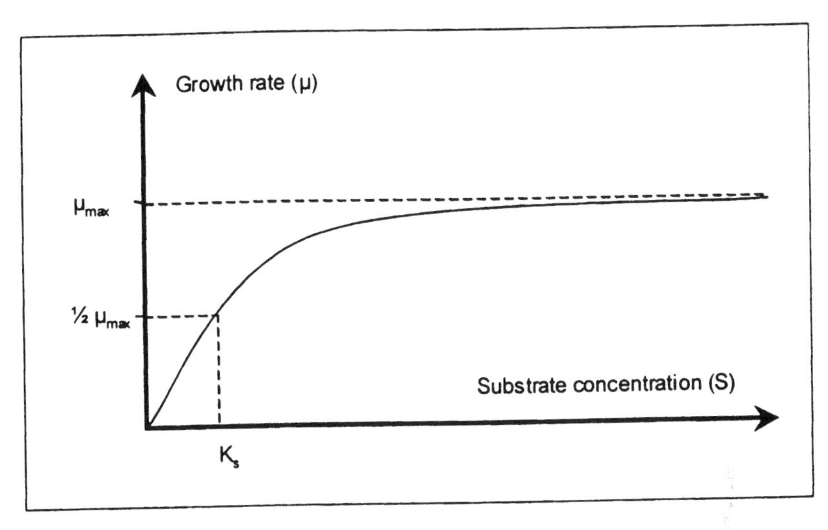

Figure 5.5 Monod growth kinetics

The significance of this relationship is that the growth rate is only affected at substrate concentrations near the K_s value. At substrate concentrations higher than say five times K_s the cells have reached their maximum growth rate. Thus, increasing the substrate concentration further will not cause the cells to grow significantly faster.

A kinetic curve as the one in Figure 5.5 may be drawn for all the different compounds needed for cell division, e.g. sources of carbon, nitrogen, and phosphor. However, the growth rate will generally be limited by the concentration of only one compound. The one substrate that limits the growth rate is called the limiting substrate. Growth depending on one substrate alone is referred to as, e.g., nitrate limited or phosphorus limited growth, depending on the limiting substrate. Once the limiting substrate has been identified, it is possible to adjust the growth rate by changing the concentration of that substrate.

5.2.1 Closed systems

If a bacterial culture is sealed in a container with a limited amount of nutrients, the growth of the culture (the increase in cell number) undergoes several characteristic phases as shown in Figure 5.6. For reasons of simplification, bacterial numbers are usually represented logarithmically or exponentially. In the following, the four phases of closed system growth (also called batch) are described.

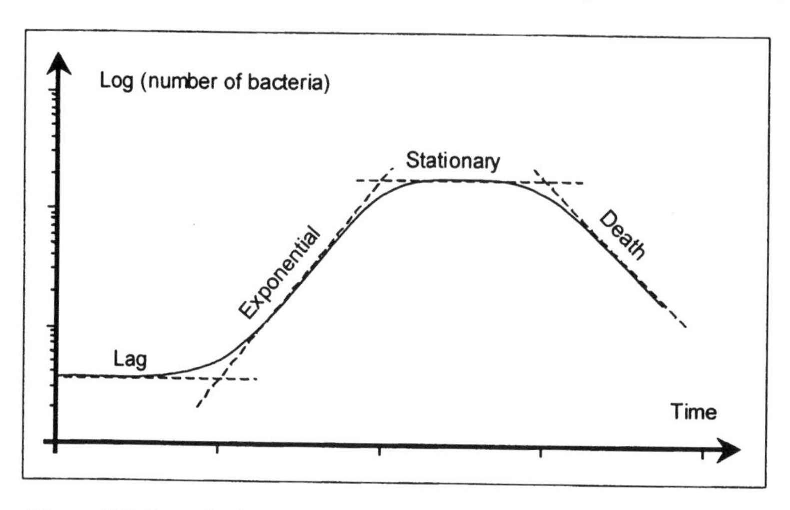

Figure 5.6 Growth phases

Lag phase

When bacteria are added to the media in the container, they may need some time to adjust to the new surroundings. They may, for example, have been inactive for a while or they may have been growing on a different substrate, and thus need to rearrange the settings of their internal codes (see section 5.1.1). Consequently, it will take some time before the bacteria start to grow at their maximum rate. This period is termed the lag phase.

Exponential phase

After having adjusted to the new environment, the cells start to grow at the maximum growth rate (depending, among other things, on the substrate concentration). Depending on the growth rate it will take a certain time for the cells to replicate themselves. Some cells may take days to divide whereas others divide in as little as 20 minutes. Every time the cells divide, the number of cells double. Thus, the total number of cells can be described as:

$$N = 2^n N_0 \qquad (5.5)$$

where n is the number of cell divisions, N is the number of cells after n cell divisions, and N_0 is the initial number of cells. This period is consequently called the exponential phase.

Imagine a single cell being added to one litre of water with unlimited substrate, within which this cell has a reproduction time of 20 minutes. The cell will divide every 20 minutes, and thus after one hour there will be eight cells in the water, after two hours 64 cells, and so on. After 15 hours the single cell will have turned into 3.5×10^{13} (35 followed by 12 zeros) cells. Since an average cell weighs about 1×10^{-13} gram, that many cells will amount to about 35 grammes of cells, which during each of the following hours ideally would be multiplied by eight. Needless to say, it would soon get rather crowded within the one litre of medium. In reality, several growth-limiting mechanisms, like e.g. the need for water and free space, will affect the growth rate long before the population reaches such numbers.

Stationary phase

In a closed system with a limited amount of substrate, the exponential growth will, at some point, reach a limit where an essential substrate is depleted. A toxic by-product reaching an inhibitory level could also set the limit for exponential growth. Having reached the limit, the number of cells within the closed system does not increase any more. At this stationary phase, the cells are not dead. There is still enough substrate within the water for the cells to stay alive, but there is not enough substrate to support an increase in cell numbers.

Death phase

Eventually, the substrate will deplete. The cells will thus no longer be able to fulfil their basal requirements for survival, and they will start to die. Like growth, the death phase is an exponential function. Consequently, in a logarithmic plot (Figure 5.6) the death phase depicts a linear decrease in the number of living cells. As the cells decay, they may serve as substrate for the cells still alive, and the decay will thus be an equilibrium between death and growth.

A technical system, like a tap water hydraulic system, may also function as a biological reactor. Consider for example a system containing 500 metres of piping with an inner diameter of 5 mm (containing about 10 litres of water), an expansion tank of 50 litres, and a number of different valves, pistons, fittings, etc that altogether contain about 2 litres of water. The total volume (V) of the hydraulic system is thus about 62 litres.

For reasons of simplification it will be assumed that the hydraulic system is completely closed.

The tap water used to fill the hydraulic system generally contains about 10^3-10^4 bacteria per ml, and the system will thus contain a total of about 10^8 (100,000,000) bacteria. Although this may seem like a large number of bacteria, this system is still quite clean.

Tap water may contain 0-20 mg organic matter per litre (as organic carbon, mgC/l). If the system was not completely cleaned before it was filled with water, the concentration of organic matter within the system after filling may be much higher, say 50-100 mg/l.

Assuming a yield coefficient (y) of 0.2, the amount of organic matter in the system, say 10 mg/l, assuming a relatively clean system, may be converted to about 2 mg bacteria per litre. The weight of a single bacterium is about 10^{-13} g and the number of bacteria within the system may thus increase with 2×10^{10} bacteria per ml or to a total of about 10^{15} bacteria within the system.

In order for the 10^8 bacteria to multiply to 10^{15}, they would have to reproduce themselves about 23 times (10^8 times 2^{23} equals 0.84×10^{15}). Assuming a doubling time of 24 hours, it only takes 23 days before the relatively few bacteria within the tap water has grown to the large numbers of 10^{15} or about 10^{10} bacteria per ml.

This example includes a number of assumptions that may be different within a real system. The doubling time and yield coefficient may vary from system to system depending on the conditions and the system may be more or less clean. Furthermore, the system could have a leak somewhere resulting in spillages, which would lead to the need for additional "make-up" water to the system. The leak could also cause further contamination from the surroundings. In such cases the system may be considered open.

5.2.2 Open systems

Microorganisms growing in an open system may, when observed from the outside, behave quite differently than organisms growing in a closed system. The open system may be fed with fresh substrate as well as microorganisms and by-products may be removed from the system. The cells within the open system may therefore stay continuously at the exponential growth phase (see Figure 5.6).

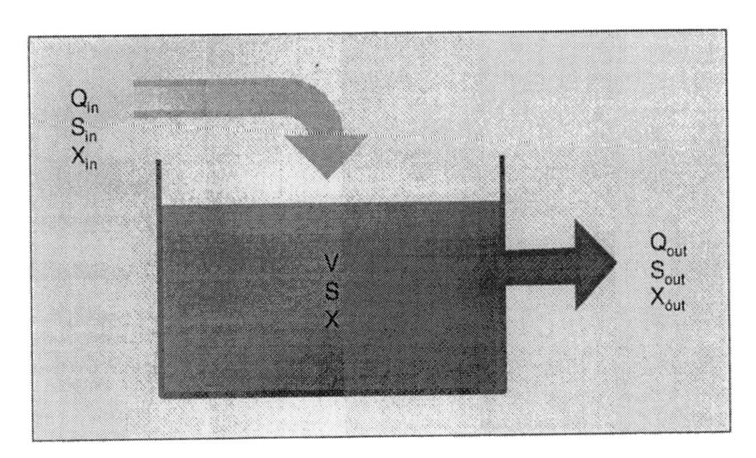

Figure 5.7 Schematic representation of growth in an open system. Q: flow, V: volume, S: substrate concentration, X: cell concentration

Consider a system with fixed boundaries like the one shown in Figure 5.7. The system is continuously fed at a rate of Q_{in} and drained at a rate of Q_{out}.

Assuming that the volume of the system (V) is constant, the flow in and out of the system must be equal.

$$Q_{in} = Q_{out} \tag{5.6}$$

Along with the flow into the system comes a substrate concentration of S_{in} and a cell concentration of X_{in}. The effluent from the system will contain cells and substrate in concentrations of X_{out} and S_{out}. If there is any microbial growth within the system, the cell concentration will increase and the substrate concentration decrease:

$$X_{out} > X_{in} \quad and \quad S_{out} < S_{in} \tag{5.7}$$

Assuming that the system is fully mixed at any time, the concentrations within the effluent will be equal to the concentrations within the system:

$$X_{out} = X \quad and \quad S_{out} = S \tag{5.8}$$

The production of cells within the system, or the development of the concentration of the cells (X) with respect to time, can be described as:

$$\left.\frac{dX}{dt}\right|_{growth} = \mu X \tag{5.9}$$

where the notation $dX/dt|_{growth}$ means the rate at which X (the cell concentration) changes with respect to t (time) due to growth, or in other words the rate of cell growth.

Analogous to the above, the flow of cells to and from the system with respect to time can respectively be described as:

$$\left.\frac{dX}{dt}\right|_{in} = DX_{in} \quad \text{and} \quad \left.\frac{dX}{dt}\right| = DX \tag{5.10}$$

where D is the dilution rate, defined as the ratio between the flow through the system and the volume of the system:

$$D = \frac{Q}{V} \tag{5.11}$$

A total balance for the concentration of cells can be formulated as the cells entering the system plus the cells produced within the system, minus the cells leaving the system:

$$\left.\frac{dX}{dt}\right|_{system} = \left.\frac{dX}{dt}\right|_{in} + \left.\frac{dX}{dt}\right|_{growth} - \left.\frac{dX}{dt}\right|_{out} \tag{5.12}$$

Assuming that the inlet to the system does not contain any bacteria, this may be simplified by:

$$\left.\frac{dX}{dt}\right|_{system} = \left.\frac{dX}{dt}\right|_{growth} - \left.\frac{dX}{dt}\right|_{out} \tag{5.13}$$

or, using the above relations (5.9) and (5.10):

$$\left.\frac{dX}{dt}\right|_{system} = \mu X - DX = X(\mu - D) \tag{5.14}$$

Applying the relationship between substrate concentration and growth rate from Monod kinetics yield (equation 5.4) the change of cell concentration in the system with respect to time as a function of substrate concentration is:

$$\left.\frac{dX}{dt}\right|_{system} = X\left(\frac{\mu_{max} S}{(K_s + S)} - D\right) \qquad (5.15)$$

In practical terms, the behaviour of this rather complex equation can be divided into three cases depending on the relation between growth rate (ℓ) and dilution rate (D):

$\mu > D$: The production rate of cells exceeds the flow of cells out of the system and the cell concentration will thus increase. Eventually, the increasing cell concentration within the system will cause the growth rate to decrease.

$\mu < D$: The flow of cells out of the system exceeds the production rate of cells and the cell concentration will thus decrease. Eventually, all cells will be washed out of the system.

$\mu = D$: The production rate of cells equals the flow of cells out of the system and the cell concentration will thus not change. The system has reached a steady state.

At steady state, where the growth rate and the dilution rate are equal and the cell concentration thus does not change, the dilution rate, or the growth rate is equal to:

$$D = \mu = \frac{\mu_{max} S}{(K_s + S)} \qquad (5.16)$$

The substrate concentration at which the system is at steady state, may thus be found as:

$$S = \frac{D K_s}{(\mu_{max} - D)} \qquad (5.17)$$

The related steady state cell concentration may be found as:

$$X = y(S \quad - S) \qquad (5.18)$$

where y is the specific yield coefficient for cell growth at the particularly limiting substrate.

In an open system, the concentrations of cells and substrate will thus depend on the dilution rate (D), the kinetic growth constants (K_s and μ_{max}), and the yield coefficient (y).

The relations described above are based on a fully mixed system, in which the microorganisms are uniformly distributed and in which the growth of each organism is not affected by the growth of others. However, as it will be explained in the following section, real systems often differ from this model behaviour.

5.3 Biofilm growth

Most bacteria do not live free floating in water (such cells are called planktonic). Instead, they are immobilised on surfaces in complex micro-communities (these cells are called sessile). Figure 5.8 shows these two types of cell existence along with an intermediate state where the cells are grouped together in flocs. In the past, the surface bound cells have been referred to as slime, dirt, and grease or similar due to the way the observer noticed them. However, the relatively new term "biofilm", which will be used in the following, generally covers all cases of surface bound microbiological growth. New insight suggests that around 90% of all microorganisms are sessile, living attached to surfaces as biofilms [Characklis and Marchall, 1990].

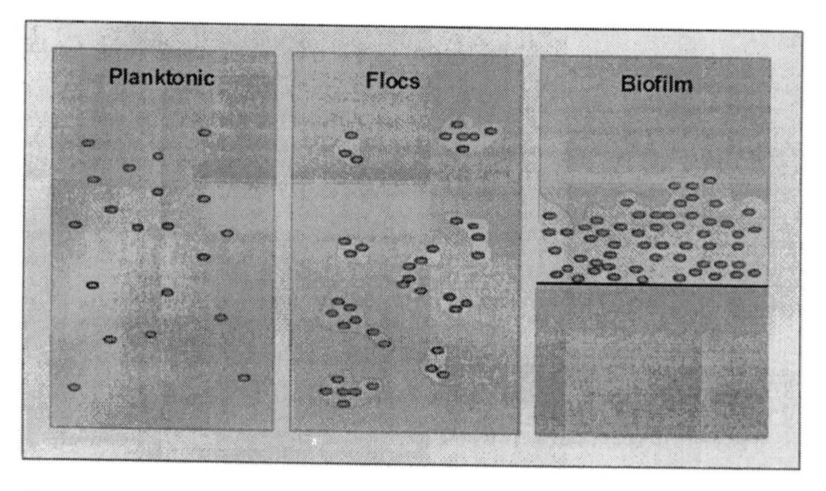

Figure 5.8 Three typical cell occurrences: planktonic (free individual cells), flocs (free groups of cells), and biofilm (cells immobilised on a surface)

Research in microbiology has largely focused on the 10% of the microorganisms living planktonically, leaving the study of sessile microbiological growth as a relatively new field that, not until the 1980s, were generally accepted by microbiologists. The existing methods of studying planktonic cells are largely based on growing the cells in the laboratory using artificial substrates. These methods have been quite limited, presumably due to the fact that only a small fraction of all microorganisms will grow under laboratory conditions (assumed to be about 1% or less). Consequently, only the part of the planktonic cells that could be grown under laboratory conditions have been thoroughly studied, resulting in quite limited knowledge of microorganisms in their natural habitat. The knowledge is rapidly growing these years, however, due to development of new in-situ methods of characterisation and identification.

5.3.1 What is a biofilm?

A biofilm is a more or less dense layer of microorganisms adhered to a surface. The thickness of biofilms normally varies from 10 to 500 µm, but film layers as thick as several centimetres have been observed. Biofilms are present on almost any wet surface, such as river stones, ship hulls, human teeth, inside pipelines, etc.

Biofilms are characterised by:

- High water content (70-95% of the total biofilm mass)
- High content of organic substances (50-90% of dry substances)
- High numbers of cells. Cell biomass constitutes 10-90% of total organic substances
- High content of carbohydrates and proteins
- (In most cases) low content of inorganic substances

When living in a biofilm, bacteria produce a gel or slime generally referred to as extracellular polymeric substances (also called EPS for short). EPS equals up to 10-80% of the total organic matter in a biofilm. The EPS consists mainly of different polysaccharides, proteins, humic substances and nucleic acids (DNA, RNA). The type and amount of EPS depends, for example, on the type of limiting substrate (e.g. nitrogen or phosphorous limitation), ionic strength and ion composition. Furthermore, the composition of the EPS probably changes with the physiological state of the bacteria (lag, exponential, etc).

One of the functions of EPS in biofilms is to keep the biofilm constituents together. EPS can further be expected to act as diffusion barriers, molecular sieves and adsorbents. The EPS is not homogeneous distributed in the biofilm matrix. The matrix contains pores, channels and areas with low concentration of

EPS. In general soluble substances can move freely within the pores and channels whereas colloidal and particulate substances are detained.

5.3.2 Formation of a biofilm

The formation of a biofilm takes place in three phases. Figure 5.9 shows a schematic illustration of the development of a biofilm. All three phases determine the structure and function of the mature biofilm and all three phases are influenced by the properties of the environment in which they take place.

Initiation

The initiation phase includes 3 steps:

1. Organic material from the water adsorbs to the surface forming a conditioning film. This may happen momentarily, after the surface is immersed in water. The conditioning film can physically and/or chemically behave differently than the original clean surface.

2. Bacteria from the water adsorb to the conditioned surface. At this point the bacteria are only loosely bound to the surface (the adsorption is reversible). This initial adhesion occurs within a few hours after a surface is placed in an aqueous environment.

3. The bacteria starts to produce EPS, which enables them to stay attached to the surface (the adsorption becomes irreversible).

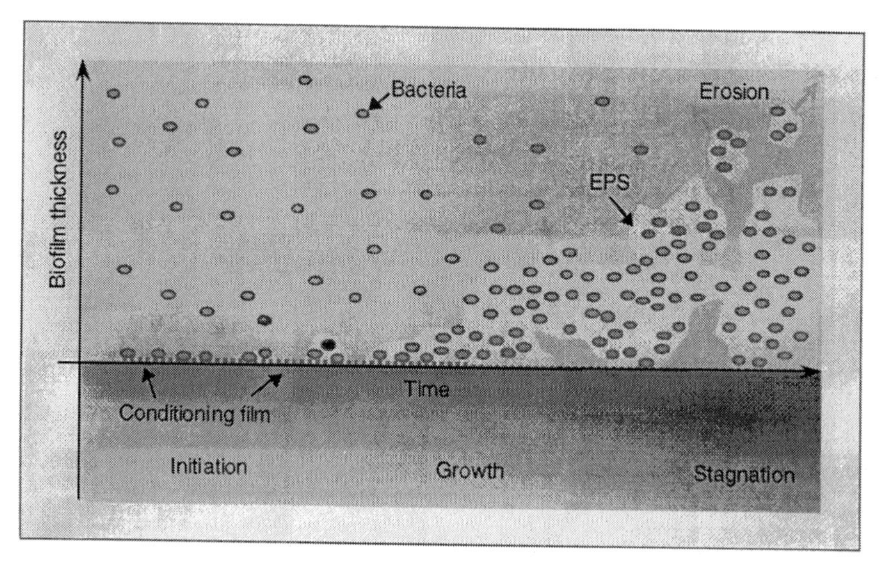

Figure 5.9 The development of a biofilm

Growth
During the growth phase, the bacteria that are attached to the surface reproduce. Substrate and nutrients diffuse from the surroundings to the surface providing food to the attached bacteria. At the same time more bacteria, along with all sorts of particles from the surroundings, might be absorbed onto the growing biofilm.

Stagnation
At a certain point the biological structure becomes so thick that the forces from the surrounding water (e.g. flow and shock waves) erode parts of the biofilm. As the cells within the biofilm continue to reproduce, the biofilm will reach a steady state where the increase due to growth equals the loss due to loosened material. This phase is called stagnation. Depending on the substrate available for growth and the external forces (e.g. flowing water), the biofilm will at stagnation reach a certain average thickness.

5.3.3 Why do bacteria grow in biofilms?

The fact that microorganisms stick to surfaces is not due to the organisms being particularly "sticky", or to put it more scientifically, having a higher affinity for the surface compared to that of the water. In fact the surface of most microorganisms are hydrophilic (they attract water) and they should therefore be repelled by hydrophobic surfaces. The glue that holds the microorganisms to the surface (the EPS) is not produced in large amounts until the organisms are positioned at the surface, so it may seem that the individual organism chooses to stay at the surface instead of returning to the water. One may thus ask the question: "why do most microorganisms grow in biofilms?" and "why wouldn't they be better off in the water being free and independent?".

Researchers have shown that by being situated in a biofilm the microorganisms obtain several advantages that they lack when living planktonically:

- Nutrients from the water are transported to the biofilm through both diffusion (transport in stagnant water driven by concentration gradients – like adding sugar to a cup of tea and waiting for it to spread) and convection (transport by water movements – like stirring the sweetened tea). In the planktonic state the organisms themselves are transported within the water and have to rely entirely on diffusion (see Figure 5.10). Though the diffusion transport to the biofilm may be of lower magnitude than to planktonic cells, the convective transport will, in most cases, still make it profitable to live in biofilms. To increase the convective transport and thus the substrate availability further, the biofilm may form internal channels enabling media to flow through the film [Stoodley et al, 1994].

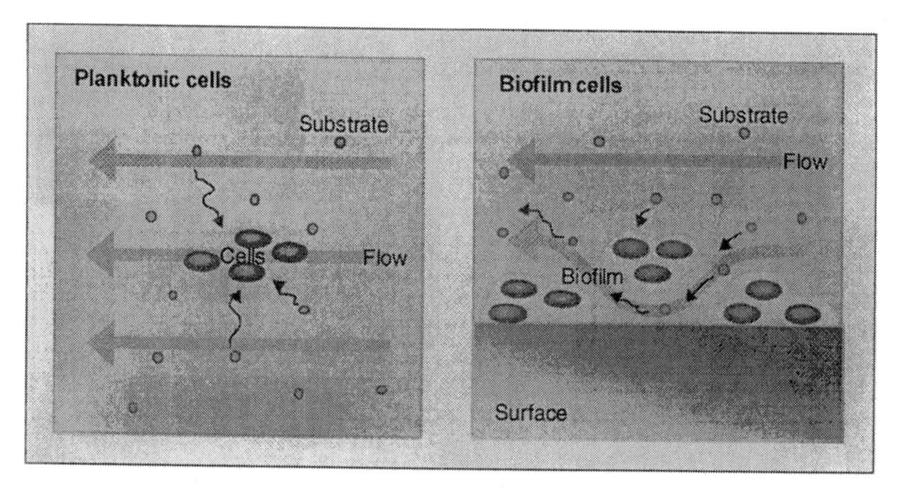

Figure 5.10 Difference in substrate supply to planktonic and biofilm cells

- The long retention time for bacteria being placed next to each other in biofilms enables development of micro consortia where different organisms perform different tasks and draw use of each other.
- Sessile bacteria in a biofilm are better protected against changes in the surrounding water (e.g. sudden changes in pH, temperature, salt concentration, nutrient situation or the presence of toxic components) than planktonic bacteria and they are much more resistant to antibacterial shock treatments.
- The possibilities for bacteria exchanging genetic material are facilitated, easing adjustments to long term external variations.

The reproduction and respiration of sessile bacteria in a biofilm creates areas with different local properties. If the water surrounding the biofilm contains oxygen, the cells within the top layer of the biofilm will use the oxygen for energy production (reduction of oxygen gives the highest energy yield). Organisms that are not capable of using oxygen will be overruled within this oxygenated layer. As one looks deeper into the biofilm there will be less and less oxygen because the bacteria above are using oxygen faster than it can diffuse from the water into the biofilm.

Figure 5.11 shows a typical profile of oxygen down through a biofilm. At a certain depth within the biofilm all the oxygen will have disappeared, even though the water surrounding the biofilm still contains the same concentration of oxygen. Below this depth the bacteria that require oxygen can not continue to grow and are thus overruled by others. If the water for example contains sulphate,

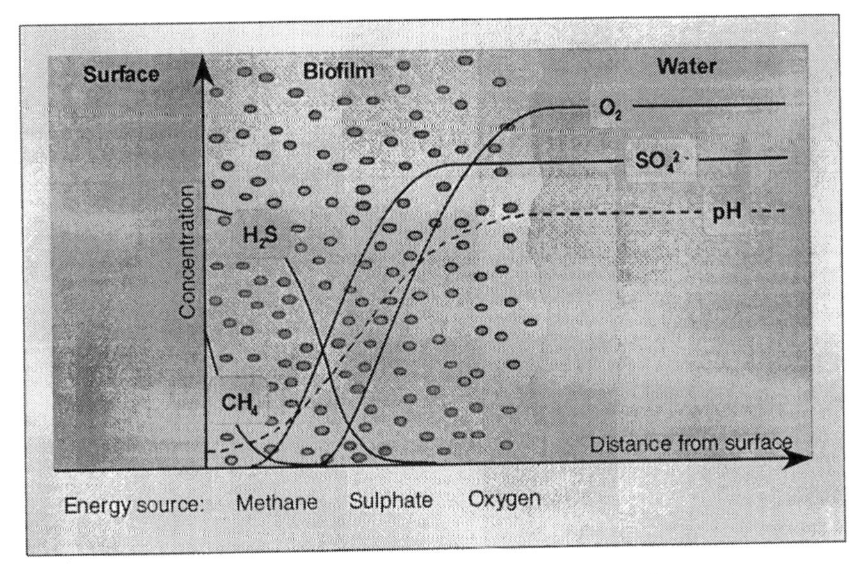

Figure 5.11 Typical distribution of different chemical variables within a biofilm

these deeper situated bacteria may reduce sulphate in order to produce energy. Moving deeper down into the biofilm, sulphate may too be depleted. At the same time the concentration of sulphide (represented as hydrogen sulphide in Figure 5.11), which is the by-product of the sulphate reduction, increases. When the sulphate is also depleted, other cells again may take over producing methane gas.

5.4 Biofouling problems

Due to the fact that the distribution of chemical compounds change with the depth in a biofilm, the chemical properties at the surface of the underlying material may be totally different that the properties in the water. The different reactions in the biofilm shown in Figure 5.11 will, for example, cause pH to drop within the biofilm. In natural waters containing oxygen, sulphate or nitrate, and organic matter, one may therefore expect to see low pH values (as low as pH 2-3 has been reported) underneath relatively thick biofilms.

Due to this ability of biofilms to change the physical properties of a surface and the chemistry near it, they may create several problems in water bearing and wet technical systems. Furthermore, the near physical presence at biofilm may in many cases cause problems on its own. Events of such unwanted biofilm formation are often referred to as "Biofouling".

As an example of the possible effects of biofouling, consider two plate heat exchangers, one fouled with a 0.03 mm thick layer of calcium carbonate scale on both sides of the plates, the other fouled with a biofilm of the same dimensions.

The calcium carbonate scale has a heat transmission coefficient of:

$2.26 - 2.93\, W/m°C$

whereas the biofilm, consisting mostly of water, has a heat transmission coefficient of:

$0.60 - 0.63\, W/m°C$

If the heat exchanger plates are made of 8 mm stainless steel, the two types of fouling will decrease the efficiency of the heat exchanger to:

- Calcium carbonate scale: 89% effectiveness
- Biofilm: 69% effectiveness

The general curve describing the relation between fouling thickness and effectiveness looks like:

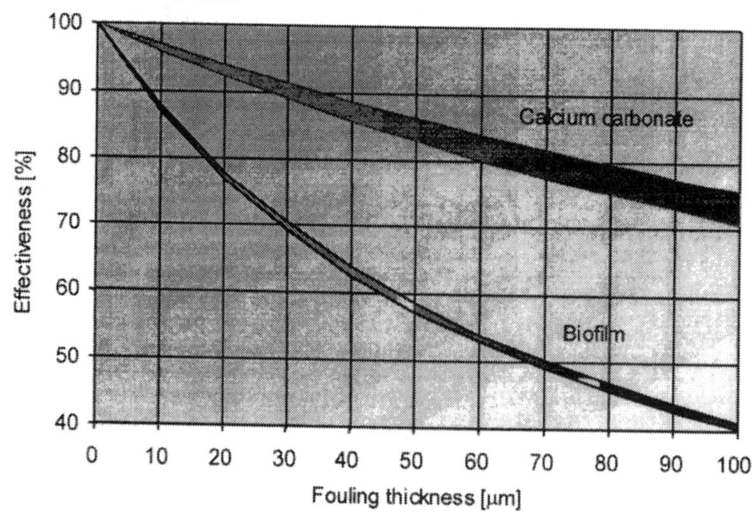

Problems caused by biofouling may be manifested in various ways, for example through:

- increased drag resistance within water bearing pipes increasing pumping costs
- clogging of filters, membranes, etc decreasing efficacy
- increased and localised corrosion damage reducing lifetime of installations
- induced changes in water quality, e.g. production of hydrogen sulphide, which may cause corrosion and health risks
- low efficacy of disinfecting agents and risk of pathogenic (decease causing) bacteria residing in biofilms.

Many biofouling problems result in loss of system performance or system malfunction. This may be the case of increased cost for pumping, heat transfer, etc. Other problems may result in induced health risks for the people directly in contact with the system or the end users of manufactured products through product contamination. Further problems, such as microbiologically influenced corrosion, degrade the system irreversibly eventually leading to leakage, breakdown, etc. In such cases, removing the biofilm that caused the initial attacks will only slow down or stop the progressing reactions. It will not solve the already existing problems of e.g. decreased pipe wall thickness.

5.4.1 Microbiologically influenced corrosion

Corrosion of metals, and for that matter degradation of other materials too, is a process closely related to the environment in which the material is situated. Particularly the chemistry at the surface of metals determines whether or not the metal corrodes. If it does corrode, the chemistry determines how it corrodes and at which rate the corrosion takes place. The ability of biofilm cells to change the chemistry near the surface, on which they are immobilised, may thus create a potential hazard towards corrosion if they colonise a metal surface. Formation of biofilms may also degrade other surface materials, but for the purpose of this book, the following discussion will concentrate on the degradation of metals facilitated by microorganisms. More detailed descriptions of microbiological degradation of materials may be found in e.g. [Dexter, 1986], [Geesey et al, 1994] or [Heitz et al, 1996].

Microbiologically influenced corrosion may reveal itself in many ways due to the fact that the microorganisms involved may possess many different procedures for changing the surface chemistry. In the following some of the most common mechanisms are described.

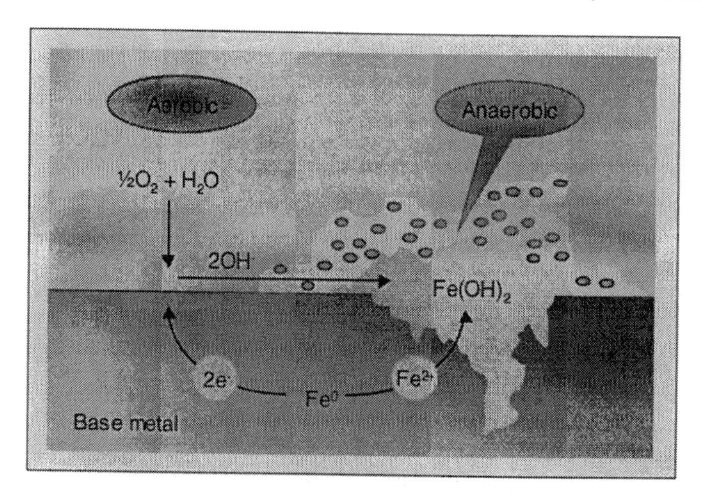

Figure 5.12 A differential aeration cell created by oxygen consuming microorganisms resulting in localised corrosion

General effect of biofilm growth

As discussed above, growth of biofilms on a surface may result in different concentrations of different chemical species throughout the biofilm. In oxygen containing water, the formation of a biofilm may cause the oxygen concentration to drop underneath the biofilm because the cells within it consume the oxygen. If the biofilm is patchy, covering only localised sites on the surface, the surface will experience an uneven distribution of the concentrations of oxygen resulting in formation of so-called differential aeration cells (see Figure 5.12).

The uneven distribution of oxygen will cause the corrosion to happen underneath the biofilm, where the concentration of oxygen is low. The areas not covered with biofilm will corrode less, but instead provide sites for cathodic reduction of oxygen (a reaction that removes the excess electrons created when the metal is dissolved). The result is that the corrosion in the oxygen containing water becomes localised instead of uniformly distributed across the surface. The overall corrosion rate, in terms of weight or average thickness lost per time, may not increase due to the localisation, but the individual corrosion attacks will be much deeper, significantly increasing the risk of e.g. through-wall pitting of water bearing pipes.

Effects of specific microorganisms

Besides the patchy effect of biofilm formation, and its tendency to localise corrosion, the biofilm may contain specific microorganisms that can make things even worse.

Several bacteria are known to produce organic or inorganic acids. *Clostridium aceticum*, for example, produces acetic acid, while some sulphur-oxidising bacteria can produce sulphuric acid. Taking place within a biofilm on a metal surface, these processes may obviously change the corrosion behaviour of many metals.

Another group of bacteria, known in relation to microbiologically influenced corrosion, is the sulphate reducing bacteria. These bacteria reduce otherwise harmless sulphate to sulphide compounds, like hydrogen sulphide, which are highly corrosive. The sulphate reducing bacteria have been shown to increase corrosion processes on mild steel as well as initiate localised corrosion of stainless steel [Lee et al, 1995], [Angell et al, 1995].

Some bacteria, like the manganese oxidising bacteria, may influence corrosion processes, even though they do not produce corrosive products [Olesen, 1998]. The manganese oxidising bacteria, for example, oxidise dissolved manganese to form insoluble manganese dioxide. On the surface of a passive metal like stainless steel, the manganese dioxide may disable the passive properties of the metal, making it susceptible to corrosion by e.g. chloride ions.

5.5 Microbial growth in tap water hydraulic systems

From a microbiologist's point of view, tap water hydraulic systems do not differ significantly from any other technical or natural systems. Bacteria and other microorganisms in tap water hydraulic systems will have the same straight-forward approach to life as microorganisms in other systems: *stay alive and reproduce*. The presence of microorganisms, active or inactive, in a tap water hydraulic system may possess problems not only to the physical performance and functionality of the system, but also to the health of workers and to the surrounding environment through leakage and spills. In order to successfully use water hydraulic systems in hygienic sensitive industries like food production, a high hygienic standard should therefore be maintained not only on the outside, but also on the inside of the system.

A tap water hydraulic system may, in terms of microbial growth, be considered either closed or open (see section 5.2.1 and 5.2.2). However, depending on the physical construction of the system, it is unlikely that such a complex system is fully mixed at any time; one of the fundamental assumptions in the above mathematical treatment of microbiological growth. Furthermore, though the system may be completely closed, the local sections throughout the system may microbiologically behave more like an open system.

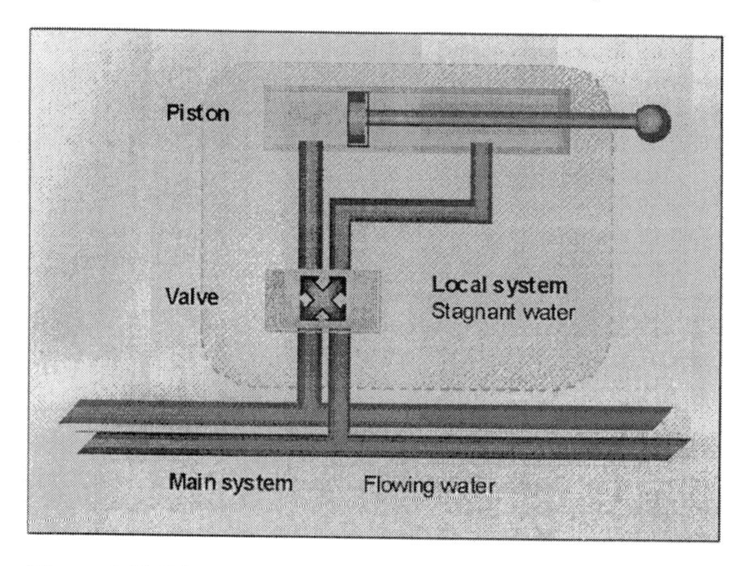

Figure 5.13 Schematic illustration showing part of a tap water hydraulic system with indications of different local conditions

Looking at a small fraction of a tap water hydraulic system (see Figure 5.13) there might be areas where the water is stagnant most of the time. This may particularly be the case around pistons etc that are not frequently used.

When not in use, the water within the local installation and within the hydraulic pipes leading to and from it will be stagnant and isolated from the rest of the system by valves etc. During such periods of time the local space limited by the valves will microbiologically behave as a closed system. The microorganisms trapped in that part of the system will continue to grow until they have used all the available substrate. Depending on the microorganisms present and the water quality, this may result in changes in e.g. pH and possibly in production of various gaseous compounds.

When the valve is eventually opened and the piston is moved, the water within the previously closed part of the system becomes more or less mixed with the water within the main supply pipes. However, depending on the volume of the water used to move the piston and the volume of the water within the pipes connecting the piston with the main system, not all the water within the local system may be replaced. It is likely that biofilm will form on the inner surfaces of the local system, but the periodically stagnant conditions will also facilitate the growth of planktonic cells.

In the distribution pipes the situation may be quite different. Here the water may be pumped around in the entire system at a high flow rate, resulting in a very high dilution rate for the individual sections of the pipe. The high dilution rate will cause the planktonic cells to be transported through the individual sections (see section 5.2.2).

Formation of biofilms within the distribution pipes and other parts of the tap water hydraulic system with a high dilution rate may be expected. In order to minimise biofilm formation in these parts of the system, the flow velocity should be kept constantly high (1.5-2 m/sec). In that case most cells will not be able to secure themselves to the surface before the high drag force from the water removes them.

The effort will, however, have no effect if the flow rate is lowered periodically, for example if the entire system is shut down during weekends. In that case it will be possible for some cells to secure themselves sufficiently to the surface during these periods enabling them to resist the high drag forces during subsequent periods of operation.

Looking at the tap water hydraulic system as a whole, and assuming that it is completely closed, the overall development in microbiological growth will have to follow the rules applicable for closed systems in general (see section 5.2.1). If no substrate is added to the system, one would expect to see an increase in the number of cells within the system during the first period of operation. The initial period would be followed by a period of stagnation, starting when the substrate in the water is running out and a period of decay when all the substrate has been used. The amount of bacteria within the system will thus, if the system is closed, depend only on the initial water quality and cleanliness of the system.

This phenomenon is the onset for a very simple, yet very effective, strategy (described in the following) to control the microbiological quality of water within such systems: keep the system closed and clean. Eventually, no matter what physical/chemical conditions exist in the tap water, the microbiological growth will be lowered/eliminated due to lack of one or more substrates.

The following describes the monitoring of microbial growth in a pilot tap water hydraulic system [Frølund and Nielsen, 1999].

Initial cleaning: 1) Tap water for 20 minutes. 2) A water-based cleaning agent (-2% w/w Ren-cid®, containing detergents, phosphate, and a complexing agent) for 2 hours. 3) Tap water for 30 minutes. 4) A disinfectant (-2% w/w Chlori-cid® containing citric acid, sodium chlorite and sodium carbonate) for 30 minutes. 5) Tap water for 30 minutes. In-between each solution the system was completely drained. Finally the filter was changed and the system filled with tap water.

The system was then run continuously at 35-40°C for three months, during which the microbiological water quality was followed. The results (below) showed growth in the beginning of the experiment (day 0 to 10) followed by starvation and death (after day 10) resembling the behaviour of microbial growth in a closed system (see section 5.2.1).

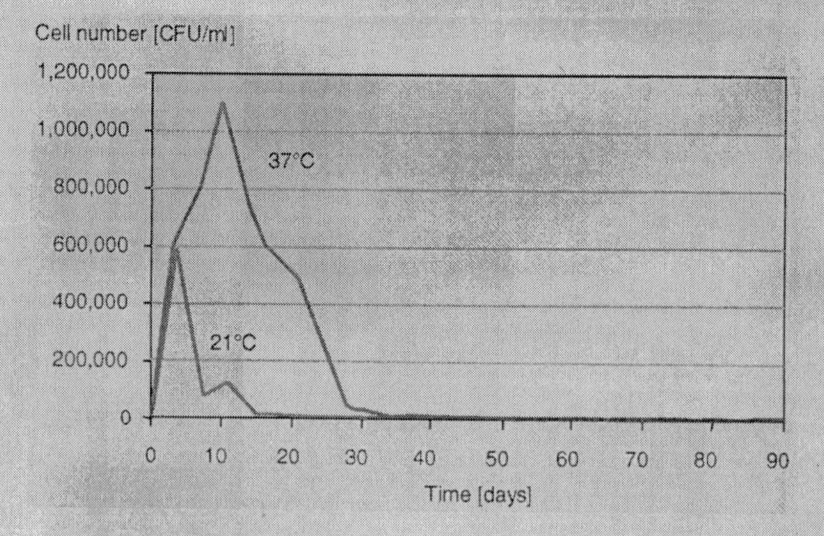

The fact that the planctonic cells vanished may not necessarily mean that the system was free of microorganisms. If a biofilm developed in the system the cells in the film would not have shown on this type of water quality analysis.

5.6 Controlling microbial growth

5.6.1 Preventive actions

One might see the optimal prevention of microbial growth in a tap water hydraulic system as sterilising the entire system and filling it with sterile water. However, as discussed earlier, the actions required to create and maintain a completely sterile system are beyond the possible economic limits for most installations. On the contrary, actions should be taken in order to control the development of the microorganisms that evidently will exist in the system.

For a given system a set of maximum acceptable levels (see Figure 5.14) can be set for all relevant water quality parameters including microbiological variables. All the different aspects of microbiological growth in the system has to be considered in order to determine how much growth can be accepted in the particular system. Having done so, the following exercises include monitoring the microbiological water quality (see section 5.7), taking the necessary actions in order to keep the variables at acceptable levels.

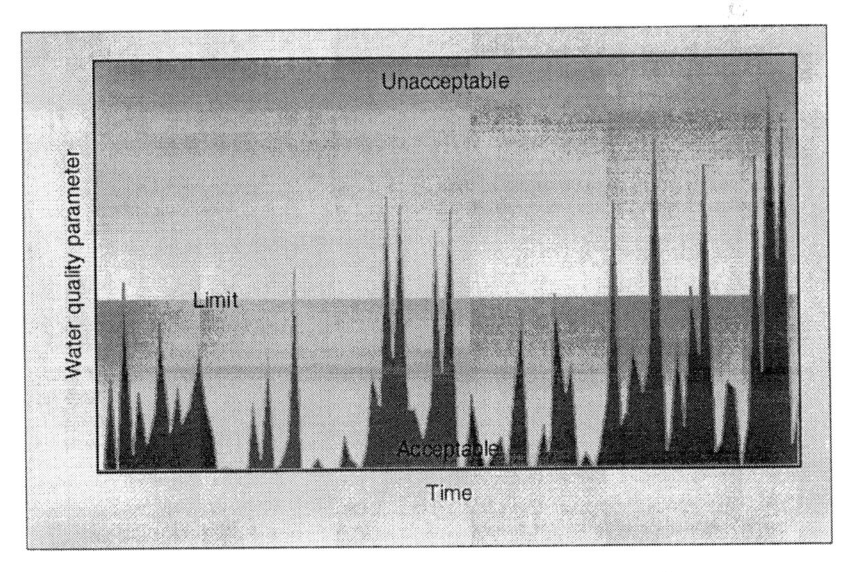

Figure 5.14 Illustration of any particular water quality parameter with indication of a maximum acceptable level

Although factors like temperature, pH and redox potential may influence the type of microorganisms present in water, the abundance and availability of an energy source (substrate) generally limits the extent of the microbial growth. Heterotrophic bacteria are the most common organisms in water, and for these organisms reduced organic compounds often represent the limiting energy source available. Many bacteria can though survive at low concentrations of organic carbon [Mueller, 1996] particularly in open systems continuously fed with fresh water. Problems associated with microbial growth have, for example, been observed in highly purified water, which generally contains only 5 -100 μg Total Organic Carbon (TOC) per litre of water.

Nevertheless, the extent of microbial growth can be minimised by taking simple precautions concerning addition of organic matter. This is what is referred to by the phrase: "Good Housekeeping", by which basically is meant keeping a clean system. The overall strategy may have a significant effect on closed systems and can be recommended as one of several means to minimise microbial growth. Depending on the system design and the surrounding environment the microbiological water quality of open systems may also be controlled this way. Minimisation of biodegradable organic materials should always be considered and implemented to the possible extent.

Sources of organic carbon are manifold. Important sources relative to tap water hydraulic systems include:

- Tap water
- Construction materials
- The tap water hydraulic systems
- Water treatment
- Surrounding environment

Failure to address one or more of these aspects when designing, installing, and running a tap water hydraulic system may result in biofouling of the inner surfaces and planktonic growth within the hydraulic medium. In the following these sources and the possible ways of minimising them are discussed.

Tap water

The tap water used to clean, rinse and fill the tap water hydraulic system should at least fulfil the current EU's specification on drinking water. It should be ensured that the water complies with the current Directive. On the total number of microorganisms in drinking water, for example, EU´s drinking water Directive, 80/778/EEC (1980), recommends a maximum of 10 cells/ml at 37°C and 100 cells/ml at 22°C.

Construction materials

Organic substances may migrate from the materials used to construct the hydraulic system and serve as substrate for microorganisms. Therefore, attention should be paid to this subject when choosing construction materials. Particular attention should be paid to the use of polymeric materials like plastic and rubber.

The most commonly used polymeric materials cover polyolefines (polyethylene, polypropylene), polyvinylchloride (PVC), polystyrene and polyester. In addition, a long list of more specific polymeric materials is used for different construction purposes including acrylonitril-butadien-styrene terpolymers (ABS), polyamides (PA) and polyurethanes (PUR).

If the use of polymeric materials is required, it is important to chose types that are "environmentally foreign" instead of "-friendly". The "environmentally friendly" products may be biologically degradable and may thus constitute a food source for microbiological growth. It is also important to recognise that many polymeric materials contain chemical additives, like softeners and stabilisers that may serve as substrates for microorganisms too. Another important aspect of using polymeric materials is that these may be degraded by cleaning and disinfecting chemicals by some of the welding techniques used. Such degradations may release biodegradable compounds from otherwise stable materials.

The physical design of components may also influence the development of microbial biofilms. Rough surfaces as well as corners of more or less stagnant water will support formation of biofilms. Furthermore, areas with complex geometry will be difficult to clean both for residual conserving substances and for established biofilms. The ideal component design thus has to include these aspects reaching for smooth inner surfaces and simple geometry with soft curves. Some of these considerations are probably already included in the component design in order to minimise friction, pressure loss and pumping expenses rather than to minimise biofouling.

The tap water hydraulic system itself

The components used to construct a tap water hydraulic system may be preserved with oil, grease, or various chemical compounds like e.g. glycol, in order to disable corrosion during transport and installation. If these substances are left in the system, they may serve as substrates for microbial growth. In that case, the system may be filled with high purity water and still experience biofouling problems. Obviously, it should thus be ensured that the entire hydraulic system is cleaned in the best possible way before it is filled with water.

The use of grease in a tap water hydraulic system may serve as a source for organic matter and thus as substrate for microbial growth. The following example describes the growth of microorganisms in closed containers containing four different types of grease [Frølund and Nielsen, 1999].

A thin layer of grease was added to 50% of the internal surface area of a test bottle. Cold tap water was added. A reference without grease was included. Bottles were stored at 25–28oC and were gently shaken. Samples for analysis of cell numbers (CFU) were taken weekly during a 12 week period. The results (below) showed about 100 times more cells in the flasks with grease compared to the reference. One of the greased flasks showed a decline in the number of planktonic cells after 8 weeks. This could perhaps be an effect of increased biofilm growth.

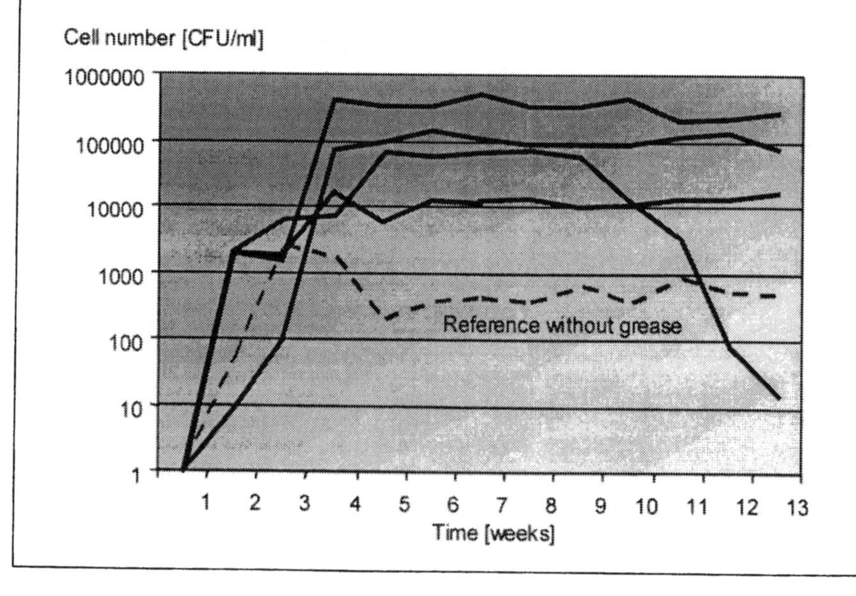

The cleaning should be carried out when the installation is complete, but before the system has been brought into operation. A preliminary pressure test involving filling the system with water and pressurising it should be performed after the system has been cleaned using water of the same high quality that eventually intended within the system. If, for some reason, the pressure test is performed before cleaning of the system has taken place, it is essential that the system is cleaned immediately following the test. Otherwise, the mixture of conserving chemicals and water used for the test will enable formation of biofilms in the entire system.

The cleaning should be made by use of cleaning agent dissolving or loosening the particular substances used for component conservation efficiently, without damaging the materials used in the system itself (e.g. induced degradation of polymeric materials). An example of a cleaning procedure is given on page 152.

If, for some reason, a biofilm develops in the tap water hydraulic system, the biofilm would contain organic matter as cells, EPS, particles, etc that at some point in time may be released, increasing the organic content of the water. Killing the biofilm, in order to eliminate this problem, may be attempted, but the outcome may be disappointing. Not only may it be difficult to kill the entire biofilm instead of just the top layers of bacteria, but the dead biofilm constitutes just as large a threat to the increasing organic content of the water as the ones alive. It is therefore essential that the biofilm is removed from the inner surface of the system, in order to eliminate this potential source of organic matter. Until today, the most reliable way to remove a biofilm is by means of mechanical force, perhaps accompanied by addition of biocides, biodispersants or other chemicals.

Water treatment

In water-bearing systems, various chemicals are often added, in order to obtain specific properties within the system. In systems operating near or below the freezing point of pure water, chemicals are added to lower the freezing point. In other cases chemical compounds are added to preserve the system components from corrosion. Parts of these additives are based on organic compounds that may serve as potential substrates for microbial growth. Some additives, like several oxygen scavengers, may themselves not be particularly biodegradable, but will, through their original purpose, be degraded to compounds that can be utilised by the microorganisms.

Consequently, solving one problem might cause several others. When adding any organic compound to a tap water hydraulic system the possible influences concerning microbial growth should be a part of the overall technical and economical considerations.

One of the pitfalls frequently observed is growth caused by addition of propylene glycol. The propylene glycol appears due to insufficient cleaning of the system prior to operation. Another source of propylene glycol is when changing the water chemistry from using propylene glycol as an anti-freeze to using pure tap water. When propylene glycol is used as an anti-freeze, the concentration is so high that it inhibits microbial growth. At lower concentrations, however, the propylene glycol can act as a substrate for microbial growth. In such cases slimy biofilms have been observed and clogging of filters can be expected. A biofilm, developed under the above condition, has been characterised in one situation [Frølund and Nielsen, 1999]. Particularly on a level-watch, placed in a tank of the particular system, a thick biofilm was observed containing about 8×10^7 cells/cm^2. The water itself contained about 7×10^7 cells/ml.

Surrounding environment

The tap water hydraulic system should be installed in a manner that minimises the risk of contamination with organic matter, microorganisms, etc.

Since microorganisms basically are abundant everywhere it is difficult to keep them from entering the system. Still, simple precautions may be taken in order to minimise contamination of the hydraulic fluid. If the system, for example, includes an open expansion tank, allowing for variations in fluid volume, this will be an obvious source to contamination. If such openings to atmospheric air are necessary, entrance of particulate matter can be minimised by sealing the opening with a filter (e.g. 0.2μm). Such a filter will, however, not prevent gaseous compounds like oxygen and carbon dioxide from entering the system and changing the chemistry of the hydraulic medium. To eliminate gas entrance, the expansion tank will have to be sealed either with a controlled gas that is heavier than the atmospheric air (e.g. nitrogen) or with some kind of flexible membrane.

Contamination with organic matter should particularly be considered when the hydraulic system is operating in environments with a high organic load, like for example in food processing or waste water treatment plants. In that case the position of every component with a potential contamination risk (pistons, valves, etc) should be considered and chosen in a way that minimises the contact to organic matter. At the same time contamination of sensitive products in touch with the tap water hydraulic system may be minimised.

As an example of how important initial cleaning of a water bearing system is, this following case describes microbiologically influenced corrosion in a boiler system [Frølund et al, 1999].

The closed system delivering heat to a larger district heating network consists of boilers connected to the network through heat exchangers. The system containing about 200 m^3 of water was installed, pressure tested, and emptied. Hereafter it was left for a few months and eventually filled with very high quality purified water and taken into operation. During the next 2-3 years the water quality was followed by means of traditional chemical analysis, showing no signs of corrosive components within the water. However, the test results also showed a high initial amount of organic matter that dropped exponentially with time. After three years, an inspection of the system showed corrosive attacks on both stainless steel heat exchanger plates and carbon steel piping. The figure below shows an example of the attacks on carbon steel piping as a depth profile through a group of corrosion pits. A following investigation revealed a large amount of bacteria within the system and biofilms on the inner surfaces containing particularly corrosive bacteria [Frølund et. al, 1999]. The growth of microorganisms within the system and the resulting microbial influenced corrosion were caused by the lack of cleaning, prior to pressure testing the system.

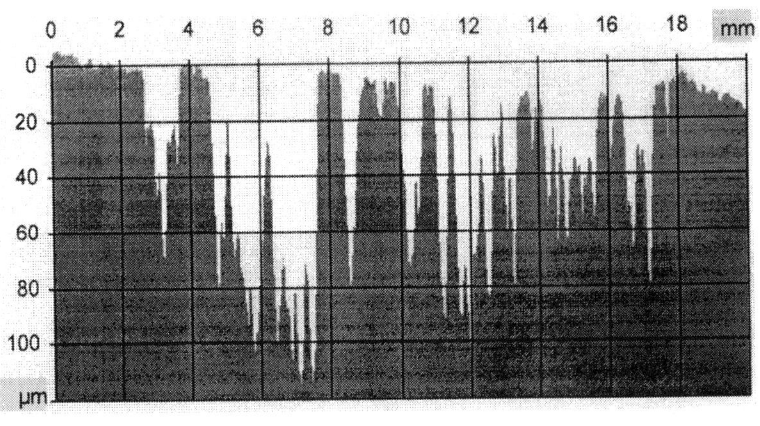

5.6.2 Mitigating microbial growth

If unwanted growth of microorganisms is discovered in a tap water hydraulic system (in the form of biofilms or planktonic cells) this should be taken care of immediately. Delaying the action of mitigation will only increase the problem.

The ideal procedure for removing microorganisms would be to take the system apart, clean all inner surfaces mechanically, put the system back together, and fill it with high quality tap water. However, this procedure may in most cases not be applicable, either due to the cost of such an action or to the system downtime associated with it, or simply because the construction of the system makes it impossible to gain access to the inner surfaces. Instead, the problem could be solved through various treatments within the installed system, possibly when it is still running. In the following some of the in situ methods will be discussed related to tap water hydraulic systems.

Thermal treatment

Raising the temperature in the system significantly above the normal operating temperature will have an inhibitory effect on microorganisms. If the normal temperature within the system is around 20°C, the culture of microorganisms within the system will consist mainly of mesophilic cells (optimum growth temperatures within 10-45°C). A periodic increase to 70°C will kill the mesophilic cells and thus the majority of all the planktonic cells in the system. If the normal operating temperature is 50-60°C the microbiological culture within the system will instead consist of thermophilic cells, capable of growing at such high temperatures. In that case it is not sufficient to increase the temperature to 70°C, since this may be within the growth range of these organisms. It will therefore be necessary to raise the temperature even more. To ease the thermal treatment and increase the efficiency hereof, it is therefore recommended to keep a relatively low operating temperature in the system.

Pasteurisation, another word for thermal killing of microorganisms, is used to preserve consumable liquids such as milk, wine, fruit juices etc During bulk pasteurisation, the liquid is held in vats at 63°-65°C for 30 min. In flash pasteurisation, which can be adapted to continuous flow yielding more satisfactory results, the liquid is heated to 71°C for 15 secs, then rapidly cooled. Flash pasteurisation is for example used to preserve many dairy products. Pasteurisation kills mesophilic bacteria and reduces the level of pathogens (organisms causing diseases). Thermal treatment can thus be recommended in case the presence of pathogenic organisms is suspected.

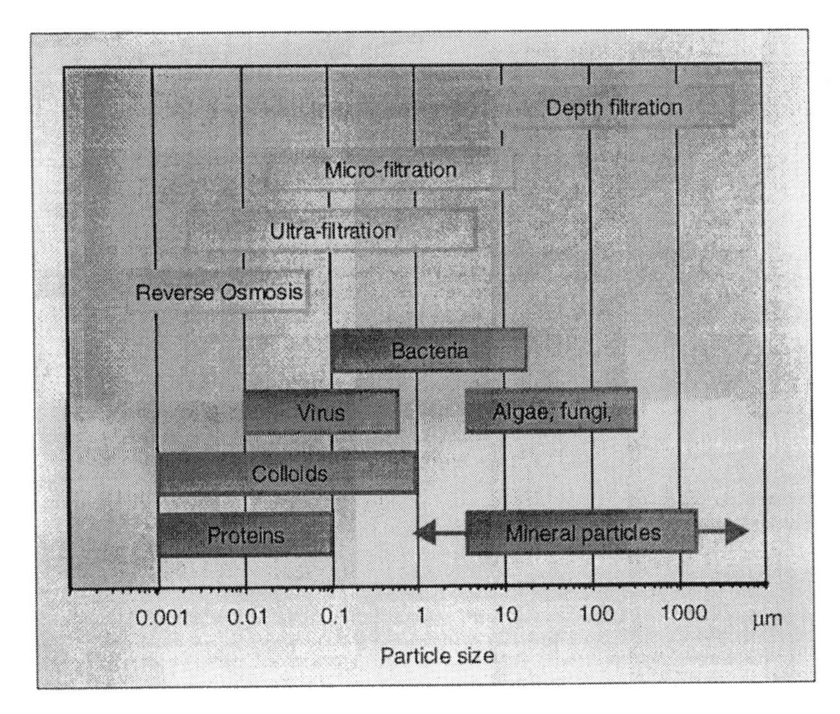

Figure 5.15 Size ranges of particles removed by different filtration techniques compared to the size ranges of various particulate matters

Before thermal treatment can be applied, two aspects have to be considered. First, the construction and design of the system should be able to withstand the temperature and the thermal expansion. Secondly, it is important to notice that the thermal treatment only kills the microorganisms, it does not remove them. The dead cells may serve as substrate for the few cells that might have survived the treatment. In this case the system may quickly be reinfected.

Filtration

Planktonic microorganisms may be removed from the hydraulic medium by filters. There are two types of filters that could be used for this purpose: depth filters and membrane filters. Figure 5.15 shows the application areas for the different filter techniques. Depth filters consist of a random array of overlapping fibre, in-between which particles may be trapped. Depth filters may to some extent remove microorganisms. Membrane filters remove particles and in some cases ions by forcing water through very small openings. Membrane filtration techniques include:

Microfiltration:	Removes particles down to micron and submicron sizes including some microorganisms.
Ultrafiltration:	Removes nearly all microorganisms.
Nanofiltration:	Removes all microorganisms and ions larger than 1 nanometer.
Reverse Osmosis:	Removes all microorganisms and most ions.

Ultraviolet light treatment

Ultraviolet light treatment is an accepted method for disinfecting drinking water. Ultraviolet light, commonly referred to as UVC, between the wavelength of 200 and 264 nanometers provides the most effective germicidal action. UV-rays with a wavelength of 253.7 nanometers are very effective as an anti-microbiological agent provided that the microorganisms are exposed to the radiation for a sufficient length of time. In terms of biofilms, UV light may not have any effect at all. Examples exist of UV lamps for water sterilisation that have been completely covered with biofilm and thus lost their effect towards planktonic organisms as well.

Ultrasonication

Ultrasonic treatment of water can be effective in inactivating planktonic growth and possibly in loosening biofilms. The principle of ultrasonic water treatment is based upon a transducer that converts a high frequency electronic signal into high frequency waves, which may shake microorganisms so vigorously that they cease to function normally. Other effects of the ultrasonic waves may be to loosen biofilm in the system and reduce formation of scale and rust. The waves may, besides the unwanted cells and deposits, also effect the system itself. Considerations concerning the use of ultrasonication should therefore also include the effect of the ultrasonic waves on the materials within the system.

Chemical treatment

Chemicals can be used in order to reduce microbial growth. However, the chemicals planned to be used in a chemical sanitation strategy must be thoroughly assessed before use. If possible, the assessment should include all aspects of the chemical e.g. legislative and technical aspects. The assessment will be a compromise between environmental, economical and other subjects. Consequently, the use of chemicals implies more than just killing bacteria. Figure 5.16 illustrates some of the important aspects.

The subjects, environment and health risk are covered by legislation (e.g. EU Directives). The remaining subjects are not covered by legislation and must be considered on another basis.

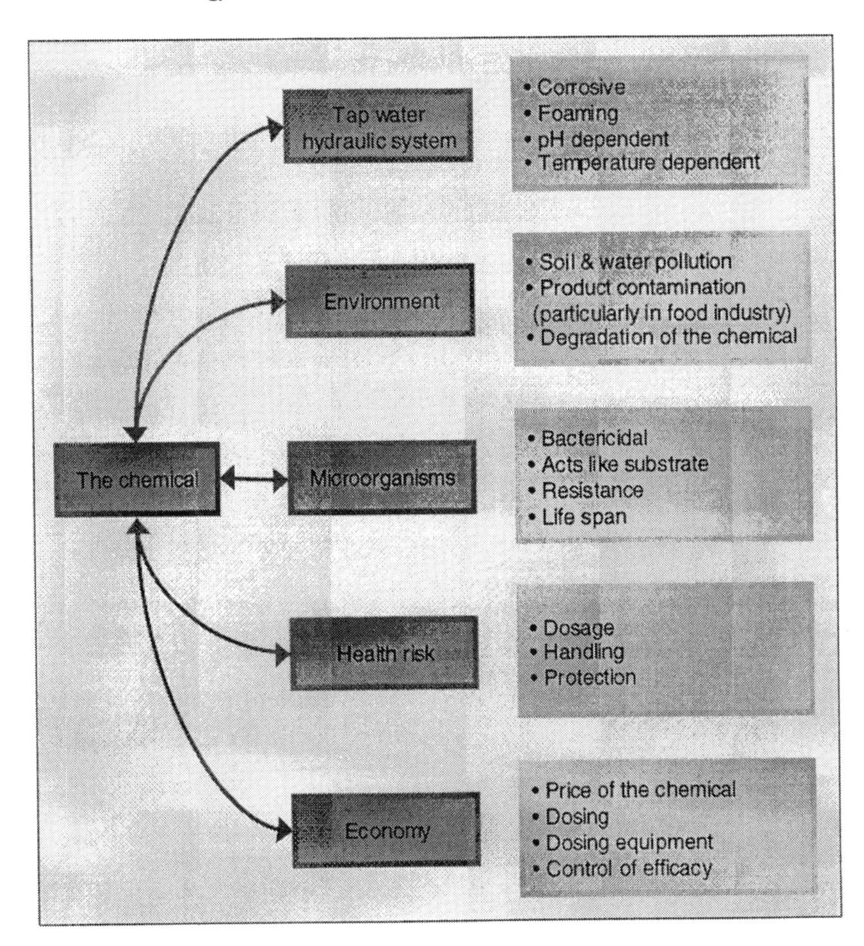

Figure 5.16 Aspects concerning planned usage of a chemical

The characterisation of biocides according to their properties is manifold. Some of the antibacterial properties are given below:

Oxidising: The biocide oxidises various chemical compounds like lipids and proteins within the microorganism surface. Organic matter in the water will reduce the effect of oxidising biocides.

Membrane active:	The biocides adsorb to the cell wall increasing the permeability and causing change in the enzyme system of the cell, finally killing it.
Electrophilically active:	The biocides react with nucleophilic substances (e.g. amino acids) at the cell membrane.
Chelate forming:	The biocides bind metal cations necessary for cell metabolism.

The antibacterial effect originating from antibacterial agents can either be bacteriostatic or bactericidal, depending on the applied concentration, the contact time, and the sensitivity of the bacteria. Bacteriostatic effects imply a reversible interruption of the bacteria activity or growth while bactericidal effects imply killing of the bacteria. The bacteriostatic effect can be neutralised either by dilution of the antibacterial agent or by addition of other chemical substances inhibiting the antibacterial agents and neutralising their effect. The interruption of bacterial growth can however not be continuously maintained. Eventually, the bacteria will either die or adapt to the biocide treatment and become resistant.

Environmental and intrinsic conditions may influence the efficiency of the antibacterial treatment. Some of these are listed below:

- A relation between the concentration of the antibacterial agent and the rate of killing has been stated:

  $$C^n \cdot t = K$$

  where K is the rate of killing, C is the concentration of antibacterial agent, n is a constant for the agents, t is the time for reaching the killing.

- The rate of killing is increased approximately exponential with increased temperature.

- pH will alter the negative surface charge on microorganisms and biofilms and may therefore particularly influence the effect of cationic biocides.

- The lack of direct contact between bacteria and the biocide is often the reason for decreased antibacterial effect. This can be caused by porous material, crevices, etc or by the targeted bacteria being situated within a biofilm.

- The solubility of the antibacterial agent in water and in the surface of the bacteria are very important factors for the antibacterial efficiency. Hydrophobic antibacterial agents are easily soluble at the bacteria surface, which increases the antibacterial effect, but they are not very soluble in water. Therefore, surfactants are often added to increase the solubility of these biocides in water.

It is well known that killing bacteria which live in biofilm is way more difficult than killing bacteria living planktonically. Resistance towards the applied antibacterial agents can develop in a bacteria population within the biofilm, if the dosage has been below the bactericidal dosage for a period. The development of resistance can be observed through faster re-growth of the biofilm bacteria after every antibacterial treatment.

The resistance can for instance be caused by:

- Modification of the structures attacked by the antibacterial agent
- Genetic change in the life functions affected by the antibacterial agent
- Change of the antibacterial agent to an inactive form.

In the following, some specific biocides are presented, describing their way of action and the implications associated with their use.

Chlorine

Chlorine effectively kills a large variety of microorganisms. Some, however, have shown to be remarkably resistant to chlorine disinfection. One of these organisms, *Cryptosporidium parvum,* was in 1993 responsible for making over 400,000 people in Milwaukee, U.S.A. sick from chlorinated drinking water.

Chlorine, being very reactive, will oxidise organic and inorganic matter alike when added to water. Therefore, not all the added chlorine will act as a disinfectant. The reaction of chlorine with water is:

$$Cl_2 + H_2O \rightleftharpoons HCl + HOCl \tag{5.19}$$

Obviously, the formation of acid affects pH. Furthermore, pH will influence the killing efficiency, due to the fact that HOCl, a weak acid, is 20-50 times more effective in its non-dissociated form than its ion OCl^-.

$$HOCl \rightleftharpoons H^+ + OCl^- \tag{5.20}$$

The ionisation constant (see chapter on water chemistry) for HOCl is 2.5×10^{-8} at 20°C.

HOCl is a non-polar molecule and it may thus rapidly penetrate the lipid structures of cell membranes. Passage of OCl^- is much slower, which is why its killing effect is so small. Although low pH is best for disinfection with chlorine, the pH is often adjusted upward to reduce the chlorine-like odour.

For chlorine to be effective, it must be present in a sufficient concentration, and it must be allowed a sufficient amount of time to react. Depending on the chlorine concentration, 30 minutes are usually sufficient for planktonic bacteria in most water systems. Following the treatment a residual chlorine concentration is sometimes maintained, in order to ensure continued protection. This may, however, induce chlorine resistance within some microorganisms resulting in the following treatments being less effective.

Excessive chloride ions, i.e. more than 200 mg/l, and free chlorine in water may lead to corrosion on stainless steel. Chlorine can combine with natural organic compounds in raw water to form some undesirable by-products. However, it does not usually pose a health problem. One concern with chlorinated water is its tendency to form trihalomethanes (THMs), carcinogenic by-products of the disinfection process.

Hydrogen peroxide

Hydrogen peroxide (H_2O_2) is an effective disinfectant, particularly when it is combined with ultraviolet (UV) light. The mechanism of this combination is based on the direct photolysis of hydrogen peroxide:

$$H_2O_2 + UV\ light \Rightarrow 2OH \cdot \tag{5.21}$$

The hydroxyl radical then participates in numerous steps, leading to the partial or complete oxidation of the organic contaminants. In addition, the irradiation may "activate" some of the organic molecules, making them more susceptible to oxidation. Concentrations of H_2O_2 greater than 100 to 200 mg/l in water are sufficient to kill microorganisms. Hydrogen peroxide is highly corrosive but environmentally safe because it is completely reduced to water through its disinfecting action. Hydrogen peroxide does, however, induce a potential health risk due to its strong oxidation action.

Ozone

Ozone is an effective disinfectant that furthermore will oxidise many organic and inorganic substances. However, it may often be a difficult substance for small systems to use. Introducing ozone to tap water would destroy planktonic microorganisms.

Ozone (O_3) is a molecule consisting of three oxygen atoms. Under ambient conditions, ozone is very unstable and as a result has a relatively short lifetime of usually less than 10 minutes. The mechanism of disinfection with ozone is directly related to the following reaction:

$$O_3 \Rightarrow O_2 + O\cdot \quad (5.22)$$

The emerging extra oxygen produces free oxygen radicals, which cause high-energy oxidation. Ozone kills bacteria by rupturing their cell walls, a process to which microorganisms cannot develop immunity. Bacterial cells viewed after ozonation appear exploded. Residual ozone concentrations of 0.4 mg/l have been shown to result in a 100% kill in 2 to 3 minutes in a biofilm, while residual concentrations of as little as 0.1 mg/l will remove 70-80% of the biofilm in a 3-hour exposure.

These properties have made ozone an effective chemical for water treatment for nearly a century. During the last 20 years, technological improvements have made smaller-scale, stand-alone commercial ozone generators both economically feasible and reliable.

Ozone is, however, quite aggressive and may lead to corrosion of components in the system. This is due to its high chemical oxidation potential. Ozone is also costly to produce both for equipment and power requirements. Besides, ozone is toxic - the Public Health Department of the U.S.A has set the maximum safe working concentration at 0.1 mg/l. It is furthermore important to note that the processes involved in production of ozone, particularly in smaller plants, may possess serious health risks.

Other biocides

Besides the traditional biocides listed above, there exist a variety of other compounds that have a similar effect on microbial growth. Some of these products, which have been used extensively in the past, are toxic or carcinogenic or provide a high danger to the environment, and the use of them has thus been prohibited. As the national and international regulations on the use of biocides develop, so do the biocides themselves. New products are constantly being discovered and tested, and the market for biocides is thus constantly evolving.

The current trend in biocide development is consequently turning to environmentally friendly products that are safe to use and to dispose of. In this regard, one should be aware that the more environmentally friendly a product becomes, the easier it is for microorganisms to utilise it as substrate or to adjust to it, becoming resistant to the treatment.

In the following a few examples of currently commercially available biocides are given. The description is concentrated on the active compound within the products, not on the products themselves, since the same active compound may be sold under many different trade names.

- Quaternary ammonium compounds may be used as non-oxidising general biocides. They adsorb to the microbial cell surface decreasing the transport of compounds to and from the cell and eventually killing it. The effect of these compounds may be more or less neutralised by anodic detergents.

- Isothiazolones are another group of chemical compounds that may be used as general biocides. These compounds also attack the cell membrane. The effect may be lowered or eliminated by presence of amines and sulphide.

- Detergents and surfactants may be used to limit the formation of biofilm. They do not posses any antibacterial function, but may break up bonds between individual cells and prevent new bonds from being made. Some detergents may be biologically degradable and should thus be avoided.

- A variety of compounds target only specific organisms, often by inhibiting some of the processes they carry out. Inhibition of sulphate reduction by molybdate ions or anthraquinones may be an example hereof.

5.7 Analytical methods in microbiology

In order to keep track of the microbial water quality within a tap water hydraulic system, or for that matter any other system, frequent analysis of some microbiological parameters is necessary. Which specific parameters are relevant will depend on the particular system and the requirements to it. A hydraulic system operating in a food production industry may for example have strict regulations regarding pathogenic microorganisms. In the following, various analytical techniques relative to tap water hydraulic systems in general will be described.

5.7.1 Handling samples for microbial analysis

As previously argued, microorganisms are present in abundant numbers practically everywhere. Thus, opening a system to take out a water sample will enable contamination of both the water sample and the system itself. As with many other analysis techniques, the methods for microbiological analysis do not differ between what was in the original sample and what was added to it during sampling and handling. It is therefore essential that samples, which are to be analysed for microbiological parameters, are handled correctly in order to achieve reliable analysis results.

The best solution would be to leave the sampling to the professionals, who are to analyse the samples. If for some reason this is not possible, one could ask for a description of how to handle samples for the particular analysis. If this too fails, one should at least take these general precautions:

- Use sterile containers/bottles for sampling (such may be retrieved from various suppliers of laboratory equipment)
- Use vinyl/latex/rubber gloves; never touch samples with bare hands
- Clean the area around the sampling site (e.g. tap or drain valve) and wipe with ethanol, prior to sampling
- Store samples cold (around 4°C).

Also, during the analysis of microbial samples, there are many potential sources of contamination. Therefore, it is recommended that the analysis is only performed by professional laboratories, in order to minimise data errors and uncertainty.

5.7.2 Total bacterial count

Normally, it is rather difficult to see individual bacteria through an ordinary microscope, especially if the bacteria are mixed with other particles. In microscopes with very high magnification, such as the scanning electron microscope, it is possible, though, to obtain quite detailed images of bacteria. However, using such high magnification it is not possible to measure the total amount of cells, since one is only looking at a very limited area. A number of methods, which have proven to be useful in determining the total amount of bacteria in a sample, are based on staining the bacteria with a chemical that emits light when exposed to UV light. The stained bacteria are easily recognised and may thus be counted at a relatively low magnification (usually 1000x). The method counts all bacteria, both alive and dead.

In order to count the bacteria in a water sample, the sample is passed through a filter with pores smaller than an average bacteria (usually 0.2 μm). The filter is then stained and the number of bacteria per square unit of the filter may be counted and related to the original volume of the filtered sample. Figure 5.17 shows a microscope image obtained in a similar way. The methods may, in some cases, also be applied directly to a surface covered with bacteria.

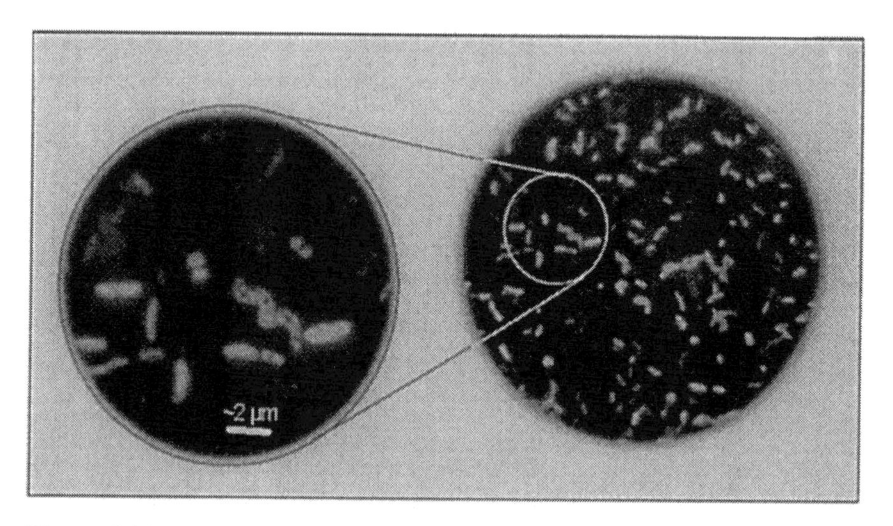

Figure 5.17 An epi-fluorescence microscope image of a filtered water sample stained with Acridine Orange. Bright spots are microorganisms

Adding formaldehyde (final concentration within the preserved sample of 2%) may preserve samples for total bacterial count. The formaldehyde kills all microorganisms within the sample and disables further growth during e.g. shipment and storage. Be aware that formaldehyde is carcinogenic and should only be handled by professionals.

5.7.3 Germination index (bacterial count)

This method is intended for measuring the number of bacteria that are alive and active. The procedure includes spreading the sample over a sterile nutrient rich gel (agar) on flat plates (petri dishes), incubating the plates for a number of days, and counting the colonies that have evolved on the surface of the gel due to microbial growth. The results are normally given in colony forming units (CFU) as the number of colonies that have evolved on the agar plate. Figure 5.18 shows an example of bacterial count on a petri dish.

The method is well known and standardised and has been used for quite some time. Unfortunately, as described in the beginning of this chapter, only a small fraction of all microorganisms can be grown on these media. Therefore, the result of such analysis does not necessarily resemble the total number of living and active microorganisms in the sample. Nevertheless, the analysed number can always be taken as a minimum content of active organisms. Depending on the specific terms, the analysis of a bacterial count may take from a few days and upwards. One should be aware that the conditions within the system from which the sample was originally taken might have changed in the meantime.

Samples for germination index can not be conserved because the cells have to be alive when the analysis is performed. Instead, the samples should be cooled and kept at around 4°C until analysis. Most methods require the samples to be analysed within 24 hours following sampling.

5.7.4 Specific bacterial strains

There exist a number of methods for testing the presence of particular organisms in a sample. These methods are normally based on techniques similar to the ones used for bacterial count. The difference lies either in the specific ingredients in the nutrient rich growth medium or in the procedure for incubation. A test for anaerobic microorganisms (organisms that can live without oxygen) will, for example, have to be incubated in an oxygen free chamber. Another example could be a test for manganese oxidising bacteria that have to be grown in a medium containing manganese. The methods may range from the simple test for a large group of organisms (like the anaerobes) to specific strains (e.g. sulphate reducing bacteria). However, the sensitive parts of all these methods lie in disabling the growth of all organisms, except the ones of interest.

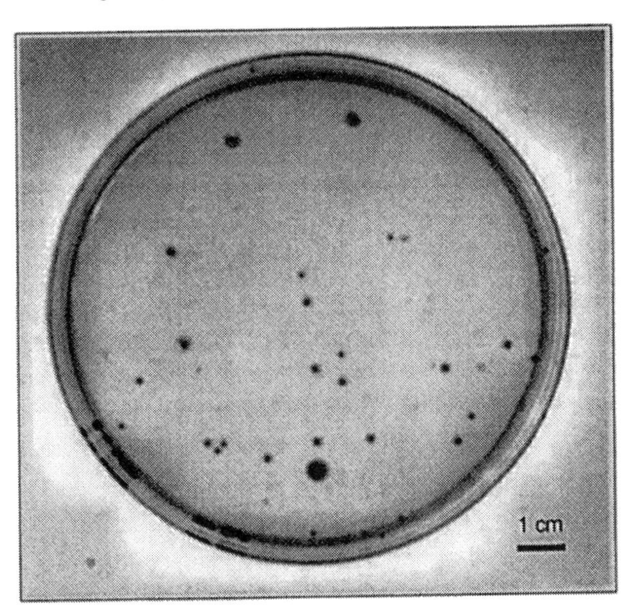

Figure 5.18 Example of bacterial count showing colonies of microorganisms on a petri dish

Instead of targeting the organisms themselves, some methods target the functions of the particular group of organisms, e.g. sulphate reduction, nitrification, iron reduction, etc.

A new era of specific microbiological analysis is evolving these years, using new molecular techniques. The new methods are based on staining techniques like the ones used for total bacterial count, using highly specialised stains that only bind to certain bacterial strains. Applied to intact samples of e.g. biofilms, these methods enable real in-situ analysis of microbiological populations, which are proving to be a very powerful tool for microbiologists in all areas.

5.7.5 Pathogenic microorganisms

The large and diverse groups of microorganisms that induce various diseases to humans are collectively called pathogenic organisms. The majority of these organisms have an optimum growth temperature around the normal human body temperature, which is why bacterial counts on drinking water etc are normally carried out at 37°C. Pathogenic bacteria in a technical system may induce a health risk to workers, particularly during maintenance and repair. If the system is in contact with sensitive products such as foods, there is also a risk that the pathogens may contaminate these products, creating a health hazard for the end user.

The analysis for pathogenic organisms are similar to the above described for specific strains. However, due to the potential danger of handling pathogens, the analysis should only be carried out at certified laboratories.

5.7.6 Indicator parameters

Even though microorganisms are invisible to the naked eye, at least in moderate concentrations, they may reveal themselves in various ways. Some organisms induce bad odours, others tinting of the water in different nuances. If the concentration of microorganisms becomes high enough, the water becomes turbid. In many cases, in fact, there is a relatively good relationship between the turbidity of a water sample and the number of microorganisms in it. Another indicator for potential presence of microorganisms is the organic matter in the water (see chapter on water chemistry). Changes in the parameters listed below could indicate growth of microorganisms and should always be followed by a more detailed microbial analysis:

- Turbidity
- Odour

- Coloration
- Unpredictable presence of gas
- Slimy/greasy surfaces
- Unpredictable corrosion
- Unpredictable clogging of filters, etc

Failure to follow up on such changes may result in further development of problems and growing cost of mitigation.

References

Angell, P., Luo, J.-S., and White, D. C., "Microbially Sustained Pitting Corrosion of 304 Stainless Steel in Anaerobic Seawater", *Corrosion Science* **37**(7), 1085 (1995).

Characklis, W. G. anf Marchall, K. C., "*Biofilms*", John Wiley & Sons, Inc., 1990.

Dexter, S. C., "Biologically Induced Corrosion", *Proceedings of the International Conference on Biologically Induced Corrosion, June 10-12, 1985, Gaithersburg, Maryland*, National Association of Corrosion Engineers, Houston Texas, 1986.

Frølund, B. and Nielsen, P. H., "Challenging the Microbiology in Closed Tap Water Hydraulic Systems", *Proceedings from The Sixth Scandinavian International Conference on Fluid Power, SICFP'99, May 26-28, 1999, Tampere, Finland, p139.*

Frølund, B., Olesen, B. H., Kjellerup, B. V., Ødum, S., and Nielsen, P. H., "Bio-corrosion in a dh peak load unit", *News from Danish Board of District Heating* **4**, 12 (1999).

Geesey, G. G., Lewandowski, Z., and Flemming, H.-C., "Biofouling and Biocorrosion in Industrial Water Systems", CRC Press, Florida, 1994.

Heitz, E., Flemming, H.-C., and Sand, W., "*Microbiologically Influenced Corrosion of Materials*", Springer, Berlin, 1996.

Lee, W., Lewandowski, Z., Nielsen, P. H., and Hamilton, W. A., "Role of Sulphate-Reducing Bacteria in Corrosion of Mild Steel: A Review", *Biofouling* **8**, 165 (1995).

Mueller, R. F., "Bacterial Transport and Colonisation in Low Nutrient Environments", *Water Research* **30**(11), 2681 (1996).

Olesen, B. H., "*Influence of Biomineralised Manganese on Corrosion Processes*", PhD Thesis, The Environmental Engineering Laboratory, Aalborg University, Aalborg University Press, 1999.

Stoodley, P., DeBeer, D., and Lewandowski, Z., "Liquid Flow in Biofilm Systems", *Applied and Environmental Microbiology* **60**(8), 2711 (1994).

6 Water treatment

Bjarne Hilbrecht
Danfoss A/S

6.1 Preface

This chapter will acquaint readers with the basic conditions for operating a water hydraulic system including water treatment methods and will also present examples of where and how the specific methods are applied. The presentation is not to be understood as a qualitative assessment of possibilites but rather as a reflection of methods and guidelines used in water hydraulic systems as of January 2000. It must, however, be underlined that the functionality of a water hydraulic system is not normally subject to water treatment. Water treatment will only be introduced on customer/application request or when required by authorities.

6.2 Good rules

As is the case for ordinary oil hydraulics, *cleanliness* is the most important rule within water hydraulics. Provided cleanliness is observed while assembling and starting the system, and the filtration requirements are observed too, an unproblematic operation will be ensured for many years. Paragraph 6.7 displays a graphic illustration of "Right and Wrong in Water Hydraulics", communicating the basic rules of design, installation and starting of water hydraulic systems through simple pictograms.

6.3 Water treatment owing to water's chemical properties

In special applications such as, for example, high pressure humidification or other processes, it has proved necessary to remove contents of salts and minerals in water. In humidification-applications, water is atomised through nozzles likely to get quickly clogged by calcium deposits. A system leakage in e.g. the textile industry must not be detectable in the fabric and thus this industry is using desalinated water, as is the case in humidification applications.

This chapter will discuss the most prevalent methods and techniques applied for softening and desalinating water.

6.3.1 Softening

A closed tank contains a resin, a polymer material with the special property of binding positively charged ions, cations. Physically, the resin may consist of a lot of small plastic balls.

Mode of operation "online"
Sodium ions are bound on the resin surface. When tap water passes the resin, calcium and magnesium ions will change place with the sodium ions binding all hardness agents to the resin.

Mode of operation "regeneration"
Once all sodium ions have changed place with calcium or magnesium ions, the system is unable to soften the water any further and will thus need regeneration. This is achieved by passing a brine through the tank making sodium ions re-occypy their original position on the resin and calcium and magnesium ions be "flushed" out of the tank.

Where/When to use softening?
Softening of water is *rarely* used in water hydraulic systems as only hardness agents are removed. Humidification applications and particularly the nuclear industry require 100% clean water, and thus desalinated, rather than softened water is applied in these applications. Water from the softening system may also

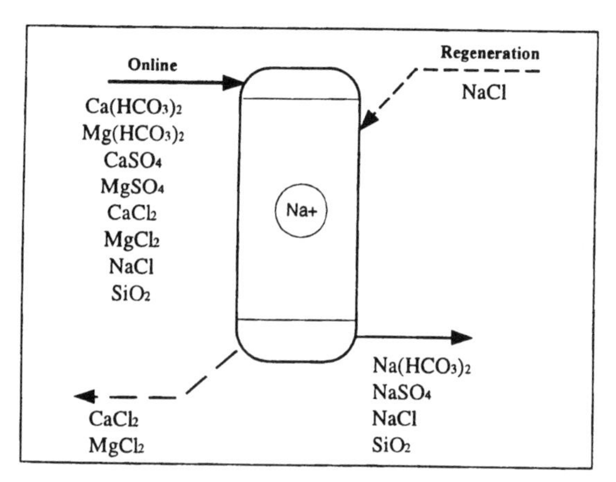

Figure 6.1 Softening through cation exchange

cause problems due to the NaCl used for regeneration. If the tank is not drained of brine after regeneration, the water filled into the system will have high content of chloride, which may cause corrosion problems when in connection with stainless steel.

6.3.2 Desalination

Desalination removes the majority of contents of minerals and salts in water. The following will discuss the two most prevalent techniques for desalinating water, i.e. deionisation and reverse osmosis.

Deionisation
Two closed tanks each contain a certain resin. In the first tank the resin property is to bind positively charged ions, cations, whereas the resin in the second tank binds anions, i.e. negatively charged ions.

Mode of operation "online"
In the first tank, hydrogen (H+) ions are bound to the resin surface. When tap water passes the resin, all cations will be replaced by the hydrogen (H+) ions. Subsequently, the water is led into the second tank, where all anions are replaced by hydroxyl ions (OH-). H+ and OH- make H_2O.

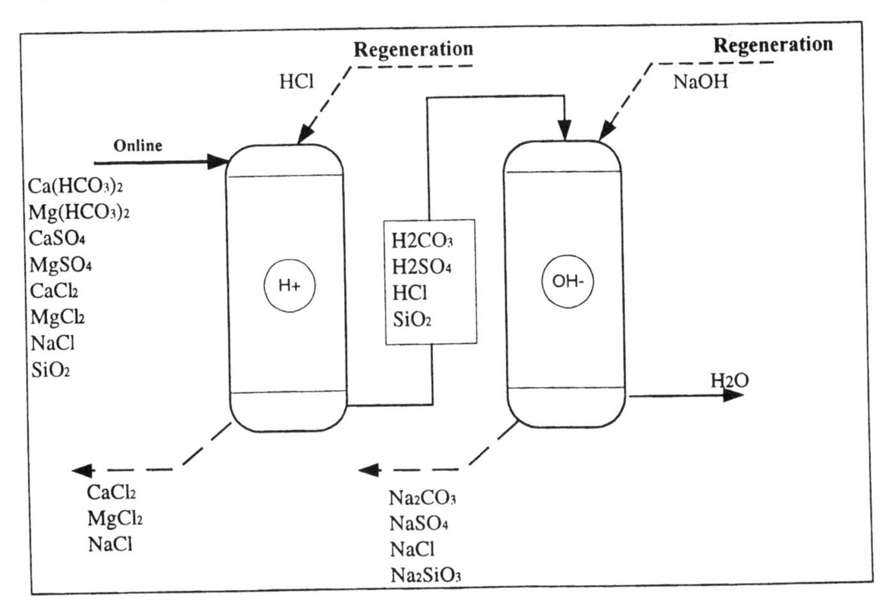

Figure 6.2 Desalinating water through deionisation

Mode of operation "regeneration"

Eventually all hydrogen and/or hydroxyl ions will be consumed from the resin in the two tanks, and the system needs regeneration. This is achieved by passing hydrochloric acid (HCl) through the first column replacing all cations bound to the resin by hydrogen ions. The other tank is flushed with caustic soda lye (NaOH) replacing all anions by hydroxyl ions.

Deionization drawbacks

Chemicals applied for regeneration require special handling and disposal. High contents of organic compounds in water will not be retained in the deionisation system but pass unhindered through. Consequently, reverse osmosis is preferred in many applications, "filtering off" organic combinations from the water. A combination of a deionisation system and e.g. ultra filtration is also frequently used.

Reverse osmosis in general

Reverse osmosis is a process filtering off contents of salts, minerals, and organic compounds in water. The filtration degree is in the nanometer area (10^{-9}m). The process is widely used in the pharmaceutic and semiconductor industry and is furthermore used for producing drinking water on seawater basis.

Mode of operation

The key element in reverse osmosis is the actual osmosis diaphragm. The osmosis diaphragm is characterised by such small "holes" that only water molecules can pass through.

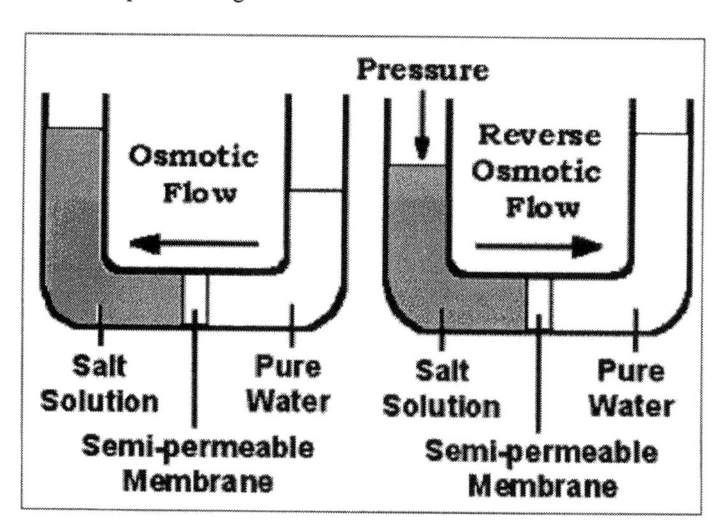

Figure 6.3 Illustration of osmosis and reverse osmosis

If an osmosis diaphragm (see Figure 6.3) was separating a container, with salt water filled in one side of the diaphragm and pure water in the other side, the water would try to balance = create identical salt content on both sides of the osmosis diaphragm. As salt cannot penetrate the diaphragm, the clean water will run over into the saline area to equalise. This process is called osmosis.

As, however, pure water, i.e. water without any minerals or salts is required, the water flow must be reversed. This is achieved by supplying pressure to the water with the high salt content, which will press the water through the osmosis diaphragm, and the diaphragm will retain the minerals and salts.

Structure in practice
If the structure shown in Figure 6.3 is used for reverse osmosis, the osmosis diaphragm would soon clog up due to the minerals and salts filtered off. In practice this is solved by letting the feed water pass over the diaphragm, also known as "crossflow". Much of the water passes through the osmosis diaphragm, but the surplus water will wash salts and minerals deposited on the diaphragm away to the sewer.

The water flushing the diaphragm has a considerably higher salt and mineral content than is the case for feed water, also called concentrate stream. From Figure 6.4 a simplified structure of the crossflow principle appears.

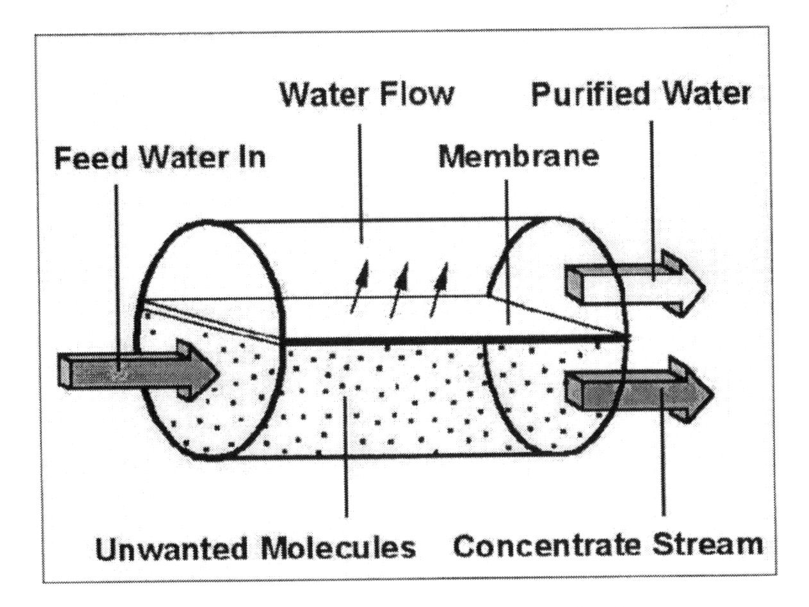

Figure 6.4 Crossflow structure of reverse osmosis diaphragm

6.4 Water treatment owing to water's microbiological properties

Keeping nutrients for micro organisms at an absolute minimum is a prerequisite for controlling the microbiological water quality in a water hydraulic system. This is typically achieved by a pre-start cleaning of the system with a soap mixture removing oil, grease, glycol and other carbon sources.

The pre-start cleaning is basis for all further water treatment (see section 6.7 for pre-start cleaning) .

What happens in water without treatment?
Provided the system has been thoroughly cleaned before filling, experience shows that a microbiological water quality similar to that of a domestic hot water tap is obtainable without any water treatment.

Figure 6.5 is a simplified illustration of the typical development of cell number in water in a closed drinking water–filled water hydraulic system. Cell number grows in a few days to 10^4 to 10^6 cfu/ml. Peak is reached approximately 2–3 weeks after the water filling, depending on system temperature and nutrients in the water. After a time, the bacteria consume all nutrients in the water and thus begin to die. Cell number will stabilise, typically after approximately 4–6 weeks on a cell number in the range of 10^3 to 10^4 cfu/ml.

To substantiate Figure 6.5, water samples were taken out at three different customers from three systems which had operated for several thousand hours without water treatment.

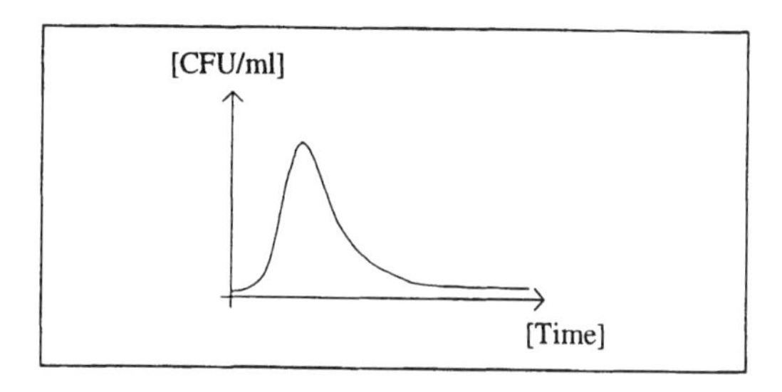

Figure 6.5 Typical development of cell number in a water hydraulic system not subject to water treatment after water is filled into the system

Table 6.1 Up-to-the minute account of microbiological water quality from different tap water hydraulic systems having operated more than 1000 hours

	CFU/ml at 21°C	CFU/ml at 37°C	TOC (mg C/l)	Temperature
TWH 1	300	4700	3.9	20-35
TWH 2	130	70	7.8	5-30
TWH 3	1200	190	2.1	5-30

TWH 1 (Tap Water Hydraulic) is a closed system used in slaughter houses. TWH 2 is a closed system used in the fish processing industry and TWH 3 a closed system used in the iron and metal producing and working industries.

6.4.1 Other methods of treating microbiological water quality

This chapter will introduce some prevalent methods providing hints on how to proceed. There are a number of water treatment methods for keeping water quality on top level, but still there is no obvious solution for covering all application areas. The system designer will typically consider the consequences of a potential leakage and select the water treatment methods meeting the demands.

The water treatment methods may be roughly split into locally working methods and system influencing methods. Locally working treatment methods are methods only affecting a certain amount of water in the system, whereas system influencing methods – as indicated by the name – influence the whole amount of water in the system.

Locally working methods

UV

A light placed inside a stainless pipe emits ultra-violet light at a wavelength of approx 254 nm (nanometer). When passing the light, the radiation will damage the genes (RNA and DNA) of the microorganisms, and thus the bacterium will be inactivated or die. UV systems are often used in the food industry.

Thermal
Part of the water flow is heated to e.g. 70°C – 80°C causing a pasteurising effect. The water in the system is max 50°C.

System influencing methods
Chemical
A number of products for water preservation are available. In the chapter on microbiology, the mode of operation for some chemicals is described, and if a chemical water treatment has been decided, there will be several questions to be asked to the chemical supplier:

- Where to use the product?
- How does it work?
- How to check the concentration?
- How and/or how often to dose it?
- What happens if a leakage occurs - is it application dependent?
- Where to buy it – water hydraulic systems are often exported to other countries?
- Compatibility with other system materials?
- Is the chemical corrosive?
- How to handle the chemicals?
- Etc.

Thermal
The water is heated to more than 70°C, whereby most of the mesophilic and psychotrophic bacteria are killed. The heat treatment of the water may be made on either a daily or a weekly basis. However, the thermal conditions in the system should be considered, as wear and the different extension coefficients of materials will have an effect at these temperatures. Thus it will always be advisable to contact the component suppliers when thermal water treatment is considered!

Combination of several methods
Combining several strategies will normally provide an improved effect and is often applied in the food industry, as a preservative may not be sufficient for taint prevention. But a combination with e.g. wrapping the food in e.g. inactivated atmosphere (protection gas) may prolong the shelf life of the product and perhaps reduce the content of preservation.

6.5 Water treatment owing to water's physical properties

As the physical state of water changes at 0°C and 100°C, respectively (at atmospheric pressure) an operational temperature limit is given. We normally prescribe the medium temperature in the system to be from 0°C to 50°C. At higher temperatures, the component service life will be reduced. At ambient temperatures below 0°C, the temperature range may be increased by adding glycol to the water. Danfoss has achieved fine experience with Mono Propylene Glycol based antifreeze having a substantially lower toxicity than ethylene based antifreeze. Mono Propylene Glycol based antifreezes are a.o. supplied from DOW=DOWCAL N and from ARCO=CHILLSAFE.

When using MPG which is easily biodegradable, it is important to note that low glycol concentrations, < 20% vol., may provoke development of biofilm in the water hydraulic system, and this will normally be obvious to the end-user by the very short service life of the return filter. If this occurs, a time-consuming system cleaning is required. In short, the reason for the biofilm development is that high glycol concentrations provide actual preserving effects, whereas low glycol concentrations act directly as bacterium nutrient (biodegradable).

"DATA"	DOWCAL N	CHILLSAFE	ASPEN TEMPER
Chemical structure	MPG: 96% Corrosion inhibitors: 2%** The rest is water.	MPG: 96% Corrosion inhibitors: 2% ** The rest is water. **Dipotassium Hydrogen phos-phate (acidity con-trolling agent from the food industry).	Salts of carboxylic-acids: 20 – 80% Water: 20 – 80% Corrosion inhibitors: < 2% Bensotriazoles: 0.3% Formaldehyde: 0.002% Sodium tetra borate: 0.05%

Cinematic viscosity **(mm²/s) at**	DOWCAL N Vol. 30%	DOWCAL N 50%	CHILLSAFE Vol. 30%	CHILLSAFE 50%	Aspen Temper -20°C	Aspen Temper -40°C
+20°C	2.7	8	2.7	8	1.82	2.89
0°C	6	18	6	18	3.41	5.8
-10°C	10	30	10	30	5.0	9.0
-20°C	Not possible	55	Not possible	55	7.7	14.9
-30°C	Not possible	120	Not possible	120	Not possible	30
	Not possible					

"DATA"	DOWCAL N	CHILLSAFE	ASPEN TEMPER
Density (g/cm^3) at **+20°C** **0°C** **-10°C** **-20°C** **-30°C**	Vol. 30% 50% 1.028 1.043 1.036 1.052 1.039 1.056 Not possible 1.060 Not possible 1.063 Not possible	Vol. 30% 50% 1.028 1.043 1.036 1.052 1.039 1.056 Not possible 1.060 Not possible 1.063 Not possible	Aspen Temper -20°C -40°C 1.15 1.23 1.16 1.23 1.16 1.24 1.16 1.24 Not possible 1.25
WGK = Toxicity grading as to ground water protec-tion	"0" Means that the agent may be used "close" to drinking water supply areas.	"0" Means that the agent may be used "close" to drinking water supply areas.	"0" May be poured down the drain acc. to the supplier. Approved by the Swedish Municipal Food Control Unit.
Fire risk	Certain conc. > 22% vol. are inflammable when atomized.	Certain conc. > 22% vol. are inflammable when atomized.	Non-flammable.
Distributors	DOW (Europe + USA)	ARCO (Europe + USA)	Sweden (Scandinavia?)
Toxicity	Classified as non-poisonous. LD50 oral rats: 21,000 til 34,000 mg/kg (MPG)	Classified as non-poisonous. LD50 oral rats: 21,000 til 34,000 mg/kg (MPG)	Classified as non-poisonous.
"NB"	Always to be mixed with DEMI. water. Residues/concen-trations below 30% may cause biofilm in the system.	Always to be mixed with DEMI. water. Residues/concen-trations below 30% may cause biofilm in the system.	Not to be mixed, but to be used concen-trated. At present we have no operation experience!

All above data are based on the specifications stated in the data sheets for the individual antifreezes.

NB: Pressure loss in systems operating on glycol and brine

The pressure loss occurring in the system due to the increased viscosity of the medium must be considered generally in all systems running on glycol.

Formulae for calculating pressure loss in smooth pipes

Reynolds numbers (dimensionless)

$$Re = \frac{Vm \cdot dH}{v} \quad (6.1)$$

Vm : medium flow velocity (m/s)
dH : the hydraulic dsiameter (m)
ν : the cinematic viscosity (m^2/s)

Resistance numbers (dimensionless)

Applies to smooth seamless hydraulic pipes.

Laminar flow => Re<2000

$$\lambda = \frac{64}{Re} \quad (6.2)$$

Turbulent flow => Re > 2000

$$\lambda = \frac{0.316}{Re^{0.25}} \quad (6.3)$$

Pressure loss in pipes

Applies to smooth seamless hydraulic pipes.

$$\Delta P = \frac{\lambda \cdot l \cdot p \cdot v^2}{d \cdot 2 \cdot 10^5} \ (Bar) \quad (6.4)$$

λ = resistance figures
d = pipe diameter (m)
p = density (kg/m^3)
l = pipe length
v = flow velocity (m/s)

Example of how viscosity and density influence the pressure loss in a water hydraulic system

Pressure loss as function of temperature and pipe dimension in pipe systems

Data : Pipe 12 x 1.5 mm => 9 mm internal diameter
Flow: 20 l/min
Flow velocity: 5.2 m/s
Result Pressure loss: bar/m

Table 6.2 Pressure loss in pipe systems 12 x 1.5 mm

(BAR/M)	+20°C	-10°C	-20°C
MPG 50% vol	0.57	0.65	1.2
MPG 30% vol	0.43	0.6	Not possible
Aspen Temper -20°C	0.43	0.58	0.63
Aspen Temper -40°C	0.52	0.7	0.9

(Water +20°C = 0.33 bar/meter)

Data : Pipe 15 x 2 mm => 11 mm internal diameter
Flow: 20 l/min
Flow velocity: 3.5 m/s
Result pressure loss: bar/m

Table 6.3 Pressure loss in pipe systems 15 x 2 mm

(BAR/M)	+20°C	-10°C	-20°C
MPG 50% vol	0.22	0.29	0.54
MPG 30% vol	0.16	0.23	Not possible
Aspen Temper -20°C	0.17	0.22	0.24
Aspen Temper -40°C	0.2	0.27	0.3

(Water +20°C = 0.12 bar/meter)

Above results prove that a larger pipe dimension is required for systems operating at temperatures below 0°C.

Pressure loss in filters
When dimensioning filters for systems running on other pressure media than water, the medium viscosity must be taken into consideration. According to Pall, the pressure loss over the filter is directly proportional to the medium dyn viscosity. At a dyn viscosity of e.g. 10 cP (P=Poise) the pressure loss will be 10 times higher than is the case for water! As a rule of thumb, the medium kinetic viscosity in cSt (centistoke) may be used as multiplier in proportion to the pressure loss in water.

However, as will appear from the list on page 1, the pressure media viscosity is factor 40(AT) to 100 (MPG) higher than that of water in worst case situations. This means that the limit values for max filter pressure loss will be quickly exceeded at low temperatures. This can only be solved by:

a) overdimensioning the filter substantially. (expensive solution)
b) using another type of filter element. Fx Pall offers filter elements with lower pressure loss
c) heating the pressure medium to decrease the viscosity. (Be aware of the return flow from the system)
d) placing a bypass valve over the filter. (No filtration at "cold starts"!)

From a financial point of view, a combination of the suggestions d and b is probably preferable.

Other related pressure media: HFC
Glycol based pressure media like e.g. HFC have been used in oil hydraulics in areas where oil is unwanted due to fire risk and environmental impacts from leaks.

Generally HFC consists of:

45% Water
25% Ethylene or propylene glycol (antrifreeze)
20% Polyglycols increasing the viscosity and with improved lubricating properties.
10% Anti foaming agents and corrosion inhibitors; some HFC compounds contain flame retardants

Comparing the properties of HFC with those for MGP based pressure media, HFC will then be characterized by:

1. e.g. HOECHST Genodyn 1791 featuring extremely high viscosity alterations, + 20°C = 120 mm^2/s and at -20°C = 3600 mm^2/s
2. being more toxic than MPG: LD 50 oral rats: 2000 mg/kg
3. by attacking certain rubber coatings

6.6 Water treatment methods within different applications

The application photos below show areas already applying water hydraulics and the type of water quality applied. The photos display a only few of the water hydraulic applications that have already been implemented.

6.6.1 Using tap water without any treatment

Picture 6.1 Water mist fire fighting equipment

Picture 6.2 Abattoir

6.6.2 Using tap water with preservative

Picture 6.3 Water hydraulic conveyor belt in high hygiene area

Picture 6.4 Aseptic filling line

6.6.3 Using desalinated water + UV water treatment

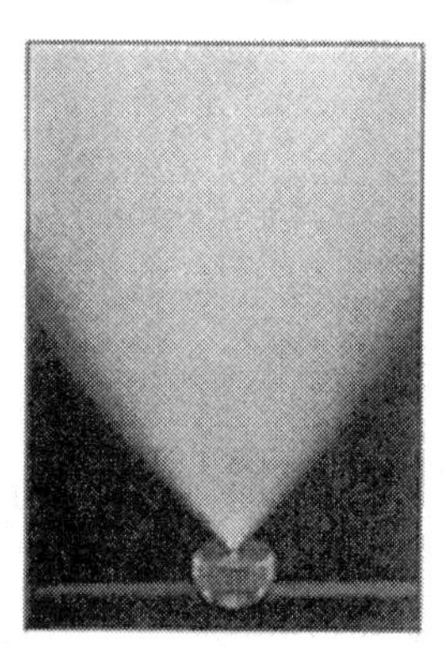

Picture 6.5 Nozzle spray in high pressure humidification

6.6.4 Using desalinated water

Picture 6.6 Paper mill

6.6.5 Using desalinated water + monopropylene glycol

Picture 6.7 Garbage truck, Sweden

Picture 6.8 Railway brake, Germany

6.6.6 Using desalinated water+borid acid+lithium hydroxide

Picture 6.9 Nuclear power plant

6.7 Simple rules for water hydraulics

Water hydraulic systems are easily handled, if a few simple rules are observed! Danfoss has gathered and communicated experience in simple pictograms. The pictograms illustrate the directions for designing, installing and starting a system.

Pictograms were chosen because pictures are stored in the human brain more easily than text.

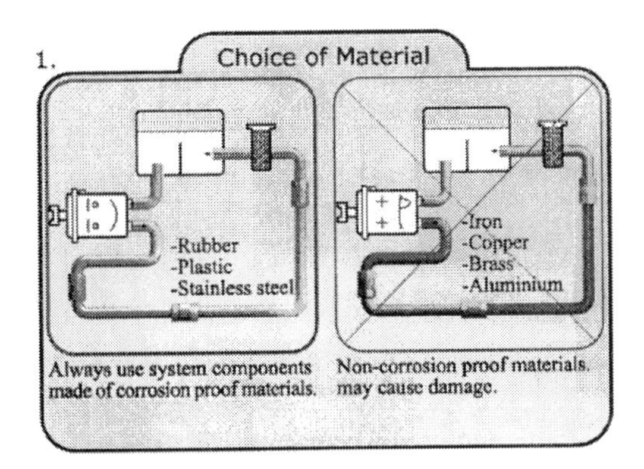

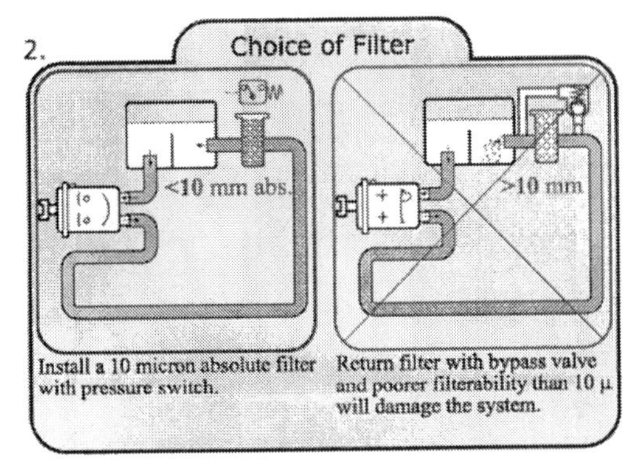

Choice of Material

Always use system components made of corrosion proof materials.
Non-corrosion proof materials may cause damage.

2.
Choice of Filter
<10 mm abs.
>10 mm
Install a 10 micron absolute filter with pressure switch.
Return filter with bypass valve and poorer filterability than 10 μ will damage the system.

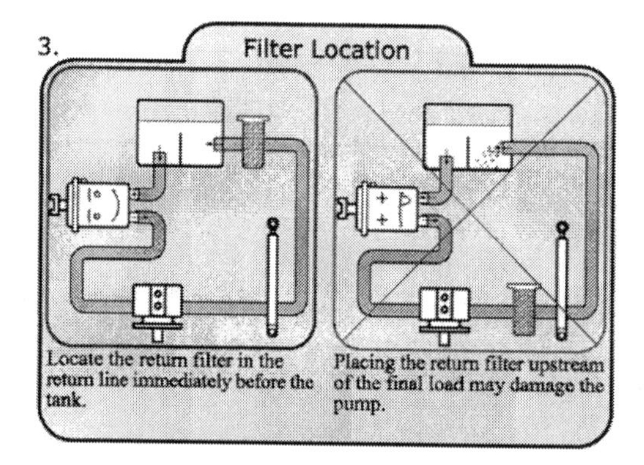
3.
Filter Location
Locate the return filter in the return line immediately before the tank.
Placing the return filter upstream of the final load may damage the pump.

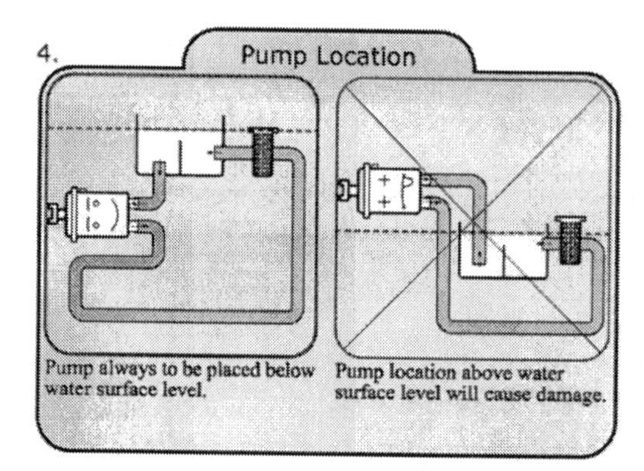
4.
Pump Location
Pump always to be placed below water surface level.
Pump location above water surface level will cause damage.

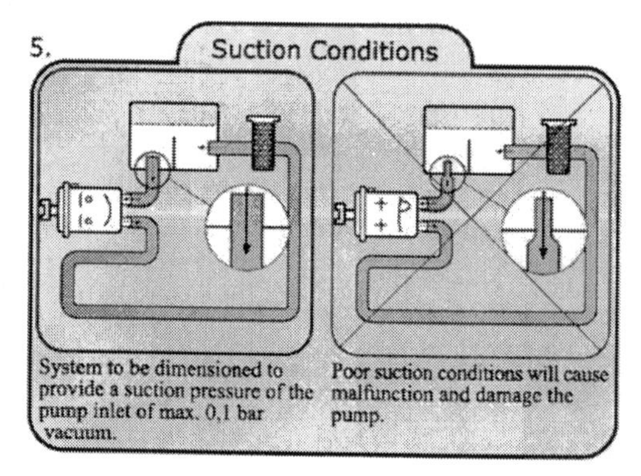
5.
Suction Conditions
System to be dimensioned to provide a suction pressure of the pump inlet of max. 0,1 bar vacuum.
Poor suction conditions will cause malfunction and damage the pump.

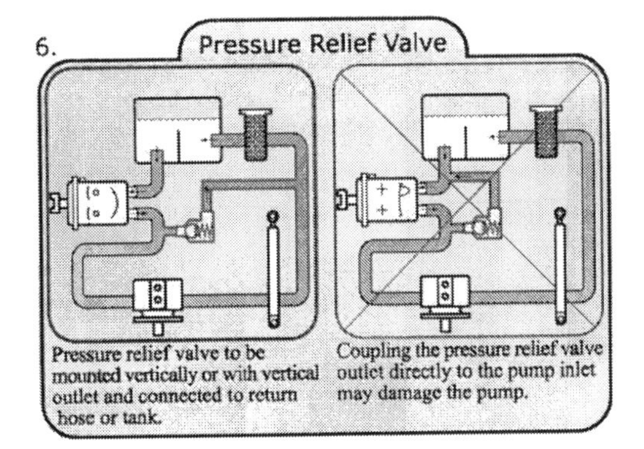
6.
Pressure Relief Valve
Pressure relief valve to be mounted vertically or with vertical outlet and connected to return hose or tank.
Coupling the pressure relief valve outlet directly to the pump inlet may damage the pump.

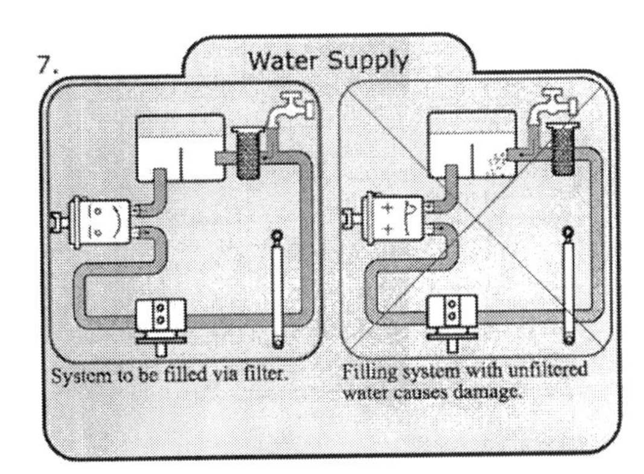
7.
Water Supply
System to be filled via filter.
Filling system with unfiltered water causes damage.

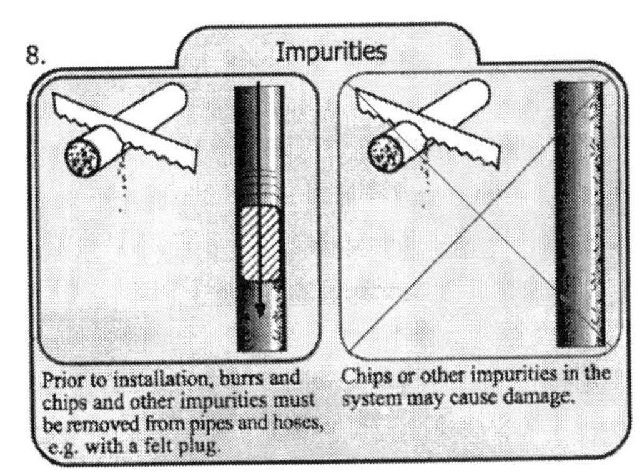
8.
Impurities
Prior to installation, burrs and chips and other impurities must be removed from pipes and hoses, e.g. with a felt plug.
Chips or other impurities in the system may cause damage.

9.

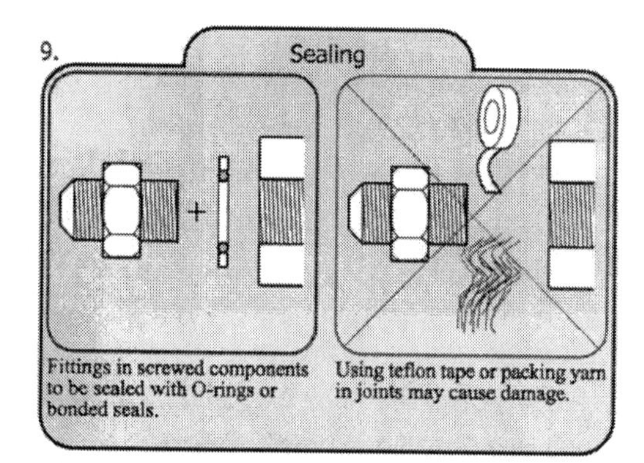
Sealing
Fittings in screwed components to be sealed with O-rings or bonded seals.
Using teflon tape or packing yarn in joints may cause damage.

10.

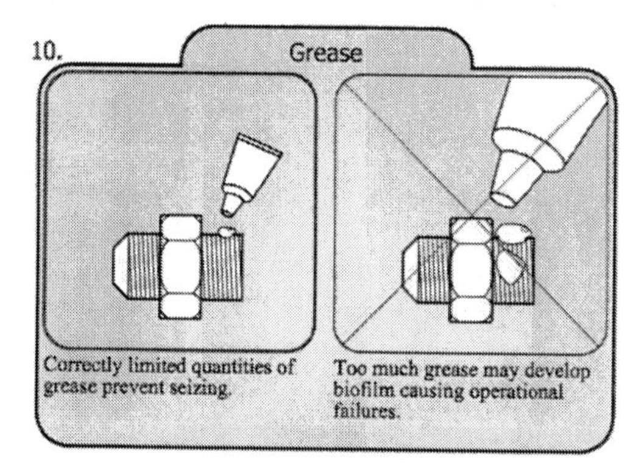
Grease
Correctly limited quantities of grease prevent seizing.
Too much grease may develop biofilm causing operational failures.

11.

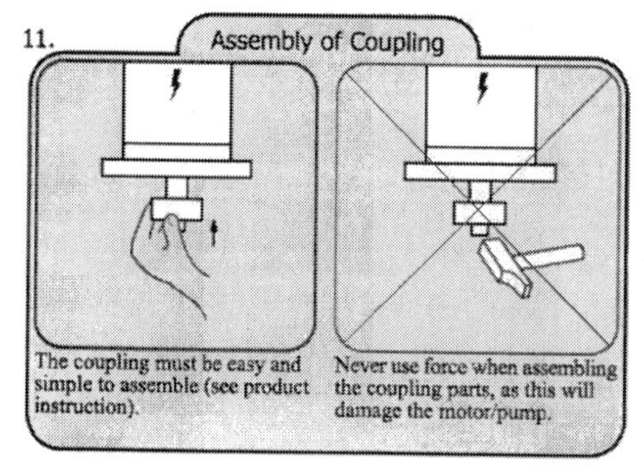
Assembly of Coupling
The coupling must be easy and simple to assemble (see product instruction).
Never use force when assembling the coupling parts, as this will damage the motor/pump.

12.

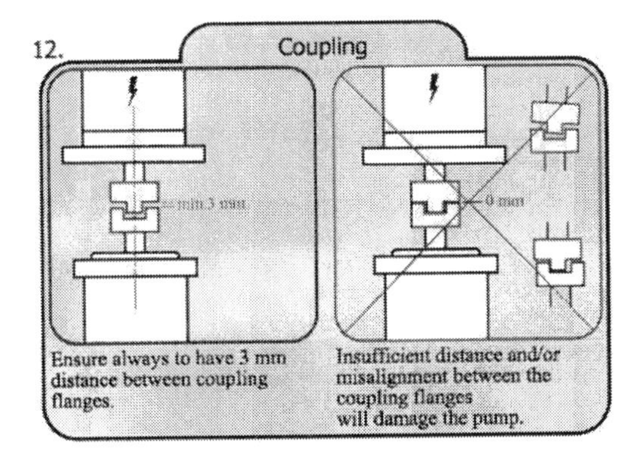
Coupling
≥ min.3 mm
0 mm
Ensure always to have 3 mm distance between coupling flanges.
Insufficient distance and/or misalignment between the coupling flanges will damage the pump.

13.

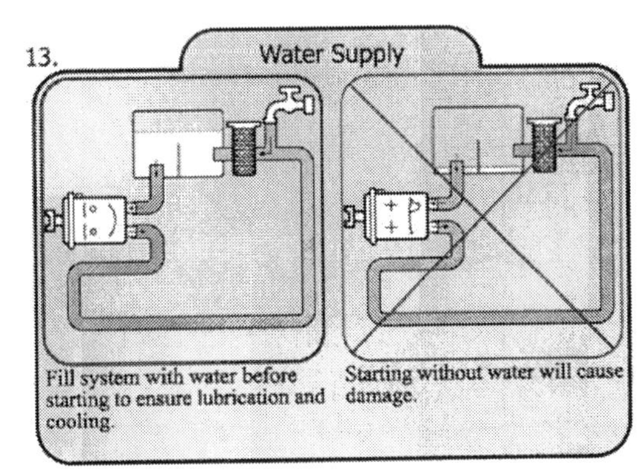
Water Supply
Fill system with water before starting to ensure lubrication and cooling.
Starting without water will cause damage.

14.

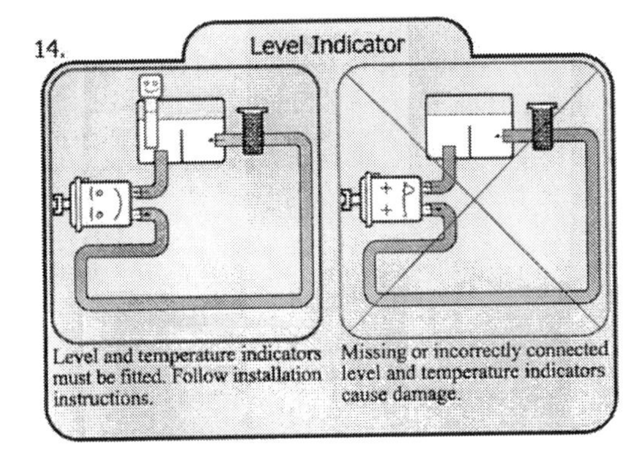
Level Indicator
Level and temperature indicators must be fitted. Follow installation instructions.
Missing or incorrectly connected level and temperature indicators cause damage.

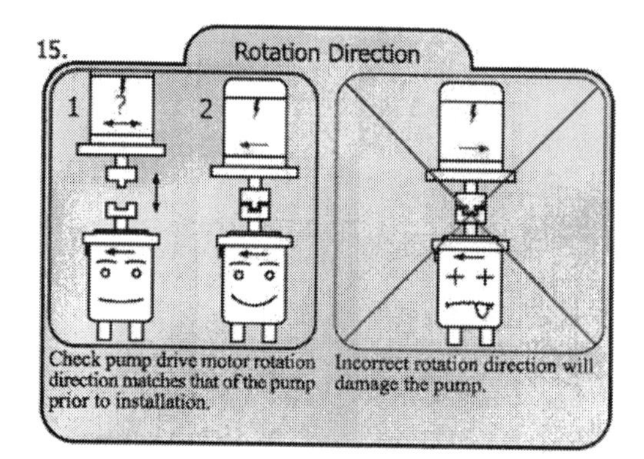
15.
Rotation Direction
1
2
Check pump drive motor rotation direction matches that of the pump prior to installation.
Incorrect rotation direction will damage the pump.

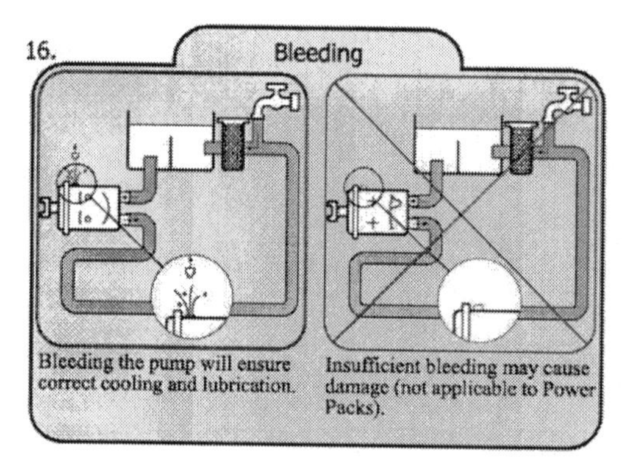
16.
Bleeding
Bleeding the pump will ensure correct cooling and lubrication.
Insufficient bleeding may cause damage (not applicable to Power Packs).

Starting Procedure

Cleaning procedure

1. Fill cold water into the system via the return filter and bleed the pump. (Power Packs PPH 4-6.3-10 and 12.5 are self-bleeding).
2. Start and bleed the system - without pressure by opening the bypass valve.
3. Add the cleaning agent to give 3% agent / water solution.
4. Run the system for 60 min. and activate all components as often as possible to ensure effective flushing with the cleaning agent.
5. Empty the system cleaning agent solution.

Flushing Procedure

6. Fill cold water through the return filter and bleed the pump. (Power Packs PPH 4-6.3-10 and 12.5 are self-bleeding).
7. Run the system for 30 min. and activate all components as often as possible.
8. Empty the water.
9. Alternatively the system may be flushed by running the plant without the return hose while continuously filling up water. The flushing should continue until there is no trace of cleaning agent in the return water.
10. Change the return filter element, fill cold water through the return filter and bleed the pump during start up.
11. **The system is now ready for operation.**

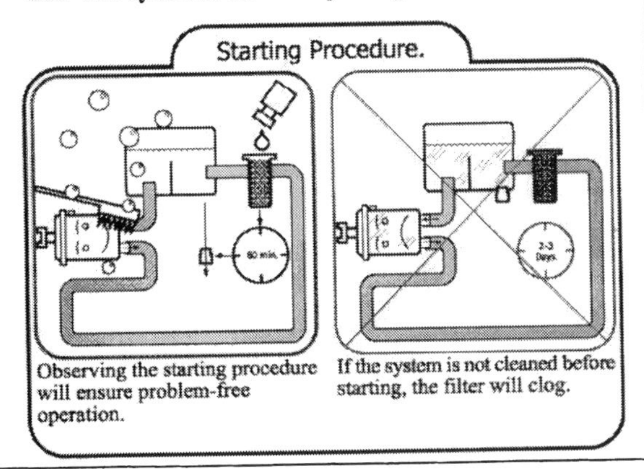

Observing the starting procedure will ensure problem-free operation.

If the system is not cleaned before starting, the filter will clog.

References

Bo Frølund and Per H. Nielsen, *"Challenging the Microbiology in Tap Water Hydraulic Systems"*, Proceedings from The Sixth Scandinavian International Conference on Fluid Power, SICFP 99, May 26-28, 1999,Tampere, Finland.

Brochures and Safety Data sheets from ARCO Chemical (CHILLSAFE)
Brochures Reverse Osmosis, Deionization from HOH A/S and EUROWATER A/S, 1999.

Brochures and Safety Data sheets from ARCO Chemical (CHILLSAFE), 1994
Brochures and Safety Data sheets from DOW (DOWCAL N), 1994
Danfoss reports regarding Environmental Friendly Glycol, 1994 and 1998

Index

A

B

C

D

E

F

G

H

R

S

T